RENATURING

Also by James Canton

Grounded: A Journey into the Landscapes of Our Ancestors
The Oak Papers
Ancient Wonderings: Journeys into Prehistoric Britain
Out of Essex: Re-imagining a Literary Landscape

RENATURING

*Small Ways
to Wild the World*

JAMES CANTON

CANONGATE

First published in Great Britain in 2025
by Canongate Books Ltd, 14 High Street, Edinburgh EH1 1TE

canongate.co.uk

1

Copyright © James Canton, 2025

The right of James Canton to be identified as the
author of this work has been asserted by him in accordance
with the Copyright, Designs and Patents Act 1988

No part of this book may be used or reproduced in any manner for
the purpose of training artificial intelligence technologies or systems.
This work is reserved from text and data mining (Article 4(3)
Directive (EU) 2019/790).

British Library Cataloguing-in-Publication Data
A catalogue record for this book is available on
request from the British Library

ISBN 978 1 83726 039 3

Typeset in Garamond MT Std by Palimpsest Book Production Ltd,
Falkirk, Stirlingshire

Printed and bound by CPI Group (UK) Ltd, Croydon CR0 4YY

The manufacturer's authorised representative in the EU for product
safety is Authorised Rep Compliance Ltd, 71 Lower Baggot Street,
Dublin D02 P593 Ireland (arccompliance.com)

To my sister, Helen

CONTENTS

BEGINNINGS   1
RENATURING   7
   On Native vs Non-Native Species   65
   On Ponds   107
   On Rewilding   133
   On the Politics of Rewilding and Wildwashing   140
   A Wildflower Field List   166
   On Biodiversity   168
   On Reintroducing Species   196
   On 'Rewilding' a Window Box   228
ENDINGS   231
ACKNOWLEDGEMENTS   243
ENDNOTES   245

BEGINNINGS

'We can never have enough of nature.'
Henry David Thoreau

I am not the owner of a large country estate. I am not a wealthy landowner. I am not a rich farmer. I moved from the city to the countryside and bought a scrap of earth. In the years that followed, I vowed that I would do what I could to help bring as much wildlife as possible back to that patch of grassy paddock. Like us all, I was worried about the climate emergency, the biodiversity crisis, the human impact on the world. I wanted to do my bit to help nature recover and flourish.

Renaturing is the story of my adventure.

Some twenty years ago, I moved to the wilds of North Essex. Behind the rather dilapidated farm labourer's cottage that I came to view with my partner was a small field just a few steps from the back door. There was a For Sale sign on the padlocked metal gate that blocked my entry to the field. I peered through the rusty wire mesh at a green expanse of rough grass.

'Is that part of the deal?' I asked the estate agent.

'No,' he said.

He knew nothing about the field.

I looked at the For Sale sign. There was a phone number. My mind whirred. A field? I could feel the fizz of adrenaline running up my body. I came from London. To have the wild open space of a field felt like the greatest freedom I could imagine.

Could we add the cost of the field on to the mortgage?

It was possible.

And so that was what we did.

Slowly, we settled into our new five-hundred-year-old cottage and the ways of rural life. For the first few years after we bought the field I had to cut the grass down every couple of weeks from early spring to late autumn. Every other Sunday I would park myself on the precarious seat of an aged sit-on mower. It had been built in the 1960s and was far better suited to the smooth fairways of the golf course for which it had been designed than the rough, irregular surface of the field. Once the engine had finally started with a cloud of grey smoke, I would drop the two cutting blades either side of me by pulling a long metal lever. We were off – careering around the field, me bumping along on my uncertain throne and leaving behind a trail of badly cut sward some four feet wide. A green, uniform patch of grass was the result – rough-and-ready and largely devoid of life.

All we used the field for was to hold an annual cricket match of sorts for friends – not that the wicket was ever anything like flat. It was a good excuse for a yearly summer party with a bonfire to sit around once the sun set. After some years, I realised that I didn't have to cut the entire field back. There was no need. I could ignore the northern and southern areas of the field and just concentrate on keeping a patch big enough for a rather ragged cricket pitch. So I would now bounce along on the mower cutting an area half the size of before.

After a few more years, a number of young oaks emerged in the longer grassy area at the far end of the field that had grown from acorns sown by jays, whose colourful brown and bright-blue underwings I would occasionally glimpse. A wilder area of the field began to be recognisable. If the ball was hit there during the cricket match, it was now considered a four. The meadow grasses in the outfield meant the older children – which, like the oaks, had also emerged and grown over the years – could entertain themselves with games of hide and seek at dusk as the adults turned to pitching tents. In the patch of the field nearer to the cottage, cars could be parked among the rising shoots of blackthorn shrubs.

Yet, once those first oaks began to appear, the entire feel of the field shifted. No longer was that patch of land merely a place of recreation for humans, it was now a natural place. Before I had bought the field it had been a large green rectangle of land, edged by fences or hedge. Within a few years of benign neglect, parts of the field were already transformed. So the seed was sown within me. After a decade and more of continual cutting to keep some form of rough playing area, I abandoned the practice entirely. The annual cricket matches were put into the long grass.

It is now ten years since I fully handed the field back to nature – more than fifteen since I unofficially began this project by not cutting the furthest reaches. Some of those young jay-sown oaklets are now twenty feet high with solid trunks six inches across. The space has been transformed. In late spring, with the oak leaves out, there is now a small copse, a secluded area of shade that the oaks have formed. One thing I have learnt is that it does not take long for a field to be dramatically changed from a fairly lifeless patch of grass to a vibrant and varied landscape.

However, the best approach in terms of increasing the biodiversity of a small plot like this is not merely to sit back and allow the field to grow as it sees fit. I needed to step in. I needed to manage the land. By encouraging the emergence of distinct areas of the field, I would enable the building of a variety of habitats which would allow for the widest possible collective of creatures and plants to live in and around that two-acre plot of land.

So I began that process of overseeing, of actively managing the field. My first task was to cut down a section of blackthorn trees which had grown up over the last decade or so, suckering across from the hedge that ran beside the green lane. Some were a couple of inches in diameter. I worked away with a bow saw. Occasionally, I was forced to halt and carefully remove one of the vicious thorns from my flesh. There was soon a question of what to do with these severed, spiky trees. I began stacking them and realised that laid horizontally they would form a rather useful dead hedge which would define an area I could concentrate on to clear and turn into a wildflower meadow. As it turned out, they would also provide shelter for a number of birds – wrens were nesting by the following spring.

I had properly begun my adventure in renaturing.

And, yes, renaturing. Not rewilding. Rewilding is about bringing nature back over vast areas of landscapes. Rewilding is an essential element in helping the globe become healthier, but it is distinctly not about small-scale human interventions, even though the word is used so often now in so many contexts. It is a common truth that the smaller the natural space to be restored, the more human intervention is required. To picture the scale for a project to truly be about rewilding, think whether apex predators can live in that space, whether wolves can roam free in that world. Rewilding needs thousands and thousands of acres. Only a handful of the largest landowners have the scope to truly rewild. Rewilding should ideally be overseen on federal, governmental scales. This is another of the truths I have learnt along the way.

RENATURING

We need to find ways in which we as individual humans can help nature recover. How can we all play our role to make more complex ecosystems so that far more variety of living beings may live happily together in the world?

We need new words.

Renaturing is one such word.

We can all renature. It is a word I came across on my journey, one I adopted and came to cherish. To renature. To actively help bring nature back to any space on the world.

If you leave some patch of your garden to be wild, you are helping bring back nature. You are renaturing. If you do not mow your lawn in May, you are enhancing the biodiversity of your garden. You are renaturing. If you are sowing pollinator-rich flowers in the window box of your flat, you are supporting the bees. You are renaturing. These are things we can all do to help make nature healthier. These are acts of renaturing, not rewilding. If we call them rewilding, we diminish the true meaning of rewilding. In the pages that follow, there are many wise voices that tell more eloquently, more emphatically of this truth.

We must never see ourselves as separate from nature. We as individual human beings are always a part of nature. We always will be. Even in the most human-constructed environment imaginable – an urban metropolis. London. Dubai. New York. Delhi. A lift in a skyscraper. Or in an aeroplane, in a submarine. Even then we are of nature.

To know this is another vital part of renaturing. In helping bring back life to any patch of land, we are also renaturing ourselves. As we act to renature, so we gently nurture ourselves, we ease that sense of eco-anxiety by actively becoming a force that is bringing about positive change to the environment on our own doorsteps.

As we shift our thoughts and think not of ourselves but of the other creatures, trees, plants that we share the space with, so we start to fall in love again with being a part of nature. In doing so we know once more our place upon this earth, feel better in our minds, stronger in our hearts, more secure in our footsteps on this world.

RENATURING

9 July
To get to the field, step from the back door of my home and pass along the tiny passageway by the brick wall of the cottage, weaving your head around long fingers of bramble and honeysuckle that now seem to grow by the day into ever longer tentacles.

You must weave, too, about the flights of bees, for there are two bumblebee colonies who live here with me and whose entrances are dark slivers of space in the masonry. Hordes of bees have been born this spring from young queens. They now fly irregular airways back and forth from their homes.

Once you have passed the bees and the bramble, you will emerge through a small wooden gate and cross the dry earth track of the ancient green lane that was here before the Romans came.

Then you step up to the field. At the entranceway you halt, for there are goldfinches feeding on the old heads of teasel, long-tailed tits flitting through the leaves of the maple trees above you. A robin watches one-eyed from the gate post. An avalanche of birdsong falls upon you.

You walk around the field on paths first formed by rabbits, deer and foxes, now forged into a human track that is always gradually growing over. You flush meadow brown butterflies from the ground before you that rise and shimmer and twist, dark flecks of shadow. And before you reach the centre of the field, two muntjac deer lift their heads and flee to the north as two white blobs on brown hides that bob and weave through the pale, high grasses. A field mouse is a brown blur upon the floor – a moment of fur briefly sighted and gone.

You step carefully on through thistle-bound regions to the fence that marks the northern edge of the field. Beyond are grass pastures shorn of life. You halt and turn and face this patch of wild, ragged earth where growth and life are so abundant, so obviously so beside the barren green deserts of surrounding lands kept only for human or horse entitlement.

The field is an experiment in wilderness. It is a glimpse, a vision of what happens when we seek to enable the earth to be as wild and alive as possible.

The field is a small patch of earth. It would seem more were it mown into an empty plain, but it has been a decade since a mower cut this grass. In the north of the field, young oaks rise from the ground. They were born of acorns from the line of mature oak trees that mark the western edge of the field where wire-mesh fencing separates the field from a reservoir of water whose grassy slopes are kept tightly shorn and virtually devoid of wildlife. This side of the fence is rather different. Blackberry bramble runs the entire length of the field. It is a mass of spiky brush, a vast wave of berries and flower and thorn that spreads some ten metres wide.

My footsteps bring me to the eastern edge of the field. It is marked by a simple three-foot-high wood-barred fence that divides the field from my neighbours' plot, where Tony and Cindy keep rescue greyhounds – up to a dozen or so at any time. They use their field as a safe space for the dogs to exercise. Each one in its own way has been damaged by humans. They care for them, whether the dog is blind or lame. It means their patch needs to

Blackberry

be mown. In the far corner their red tractor sits. Every week or two, Tony mows their acre of land as I once did mine.

I step back into the field, scattering butterflies – orange-tips and brown argus and large whites. The contrast between the two fields is stark. One is green lawn devoid of wildlife. Here, on this side of the fence, there is a hum, a constant purr – of crickets and grasshoppers, of pigeons cooing, crows cawing. Here there is a blare of colour – the vivid yellow tops of ragwort, which has done so well this year. The soft purple heads of thistles are ready to burst out. Each is topped with bugs – bees, butterflies or hoverflies – a frieze of insects.

But I am no expert naturalist. I step about the field and with each step feel my ignorance. I know the general names of some of these creatures that live here, but not their specific identities, their binomials. Crickets jump, hornets tear through the air towards me. Butterflies flutter. And I vow that in the days, weeks and months to come, I will strive to learn their true titles and their ways; and the names of the various plants and flowers that will rise from the earth of the field. And I will learn, too, to do all I can to help increase their variety and abundance on this patch of ground that I call The Field.

10 July

I ask my friend Chris Gibson to come over and guide me in my enterprise of ecological restoration. He *is* an expert naturalist. After working for Natural England for the last twenty-plus years, Chris is now newly retired.

We step into the field.

'The biodiversity will change over time,' he says.

Chris points to a hump of earth on the ground beside us.

'This is indicative of what would have been here originally – that's an ant hill growing out of the fine-leaved fescue grasses. As the coarse-leaved grasses take over, so the fescue grasses will go and probably the yellow meadow ants will go as well. Interestingly, where the yellow meadow ants are creating the barer ground is where the ragwort is settling in.'

I look about at the patch of grassland around us.

Chris explains how scattered scrub can be good for biodiversity, allowing shelter for insects. He draws a picture of the field in the future. If left for another ten years or so, a scrub of bramble and blackthorn would dominate the space, causing an overall reduction in species. In time, the trees would take over, their canopy gradually closing, shading out other growth and leading to a crash in biodiversity, which would remain low until the first trees got big enough to start falling down.

I laugh.

'Say three hundred years' time?' suggests Chris.

That is the expected 'time course' of biodiversity here in the field over the next few centuries. My mind whirrs a little at the passing of time. I look over to the cluster of young oaks already stretching out over the northern quarter of the field. Chris is also momentarily distracted.

'Song thrush,' he says.

He starts to talk about how every piece of land is different, how you need to take into account the history of the land use. That is vital. If the ground has been compacted or if a pile of compost has been left in one corner, then that will have a long-term effect on the way in which the land evolves in the future. In urban worlds, each brownfield site is unique, due to what Chris calls a 'stochastic factor' – those random elements that stem from what species of plant have blown in from neighbouring gardens or were growing there previously. Each brownfield site is essentially 'multicultural' in having both native and non-native species.

'That's what makes them so exciting to explore,' says Chris.

We turn back to the field.

'The bramble is noticeably creeping east,' I say.

I look across to the vast bank of bramble on the western edge of the field.

'A foot a year or so.'

'Yes, that sounds about right,' agrees Chris.

'It does this building thing,' I say, ineffectively describing the impressive six-foot-high wall of bramble which stands beside us.

We both stare over to the bramble as Chris starts to tell me of the way in which the nutrient levels across the country are high, how nitrogen oxides from vehicles not only damage lungs but are a fertiliser, coming into the soils through rain.

He points with a sweep across the taller grasslands of the field.

'This looks fairly fertile. You've got nettles and thistles, and creeping buttercup. All these are nitrogen guzzlers. They're doing well at the moment because we're over-fertilising – eutrophicating – the whole landscape.'

'Right,' I say, nodding my head.

The evidence is there before us. Thistles running wild, nettles doing the same in patches, and between, below, among the grasses, the yellow dots of buttercups.

We walk on over towards the young oaks.

'The best thing for nature is in fact *not* to simply leave it?' I venture.

'No,' agrees Chris. 'Certainly not in the short term and not on a small scale. It would work if you were to leave enough land for long enough and if you were to have natural dynamic factors in there.'

By that he means large creatures.

'The British landscape doesn't have them anymore. Well, it has more deer now than it had. But it hasn't got bison. We haven't got many boar. So humans have to step in and be those large creatures.'

'We have to be the boar,' I say.

It is a fascinating way of seeing things.

'So it's not negative that humans are managing land. We just have to make sure we are doing it in the right ways,' I summarise.

'Absolutely,' agrees Chris.

It is starting to make sense.

We start to imagine the site in the next few years. The bank of bramble would have crept further out. But the oaks already here would also have grown.

'Once a tree's canopy has got above the height of the bramble, it won't be adversely affected,' states Chris. 'You wouldn't get new trees generating within the bramble because they would have to grow up through two or three metres of mesh. But those oaks got in first and are developing quite happily.'

Emerging from within the vast mesh structure of the bramble

beside us are the young oaks, their juvenile branches and leaves a few feet above the surface of the bramble. They won't be drowned in the sea of spikes beneath them.

'In twenty years, these oaks will be fine, because they'll be up here,' I suggest, holding my hand out above my head, 'and the bramble will be down here.'

Chris nods and starts to explain how the next era of the field will pan out.

'By that time the canopy from the oaks will be casting enough shade to be starting to suppress the brambles anyway.'

'Really?'

'Yes, you'll start to see the bramble being reduced.'

Suddenly, I catch a glimpse of that future. Oaks tall around me. A woodland emerging. It is thrilling. I can't wait to see it.

'In thirty years?'

'In thirty years, you'll see the canopy closing, the bramble lessened, the ground will be relatively bare and the biodiversity will be depressed,' Chris pauses. 'There will be decaying leaves and suchlike, but the biodiversity will be lower until such time as the canopy starts to open again – and that can be done either by chopping it down or, as I say, waiting until the trees start to fall.'

'The trees start to do the management,' I realise.

The notion of the oaks having agency, taking control over the landscape, is striking. Yet the issue on a site like this, an old field, is that the oaks are all growing at the same rate.

'That's why there will be a biodiversity deficit – from a hundred years' time until three hundred years or so – because these trees won't yet be big enough to manage the habitat around them.'

Remove humans, step into timescales of hundreds of years, and it is now the oaks who are in charge. It is the actions of the oaks that are defining the state of the landscape.

But, of course, oaks need space as well as time. On a site like this small field, there simply isn't the scale to enable that ecological process to occur, with aged oaks hundreds of years old naturally falling to open the canopy to sunlight, and large animals roaming the forest floor, disturbing the ground and encouraging new growth. Obviously, you need vast areas of mixed-age woodland for that.

'This is where the concept of rewilding really comes in. You've got to be doing it on a seriously large scale for it to work – hundreds of hectares as a minimum,' Chris explains. 'We use coppicing to mimic the natural disturbance cycle of tree senescence. That's why I prefer the term "renaturing". You can renature on any scale.'

Renaturing.

That makes sense. Bringing back nature.

We can all renature.

I look around at the field.

We are stood still beside the young oaks which have sprung up across the northern section of the field. They are around twelve years old. Solid, strong – already way above me. There are others that are older, growing on the margins of the field. Strangely, it isn't hard now to imagine the scene in fifty years with this area all oak woodland.

It's a poignant moment to pause. We head to the cottage for coffee.

12 July

The bees are silent by the back door. I duck beneath an arbour of bramble, follow the path a few yards to the broken gate, then step over the green lane and into the field.

There is a sultry feel to the day. Summer storms came in through the night, awakening all with their clattering thunder. The dry heat has been dowsed. Moisture soaks my shoes as I step into the heart of the field. It is the fluttering of a butterfly – a large white – that catches my eye. There is always an initial something that draws you into this world of the field. It might be a sight, it might be a sound. It might be the swaying of the oak leaves, or the call of a buzzard high above. It might be the song of a robin, a spray of bright-yellow ragwort, or the soft mauve flowers in the sea of thistle that has flooded the southern section of the field and is now alive with bumblebees, honeybees, hoverflies and butterflies.

Today, the sight of the large white brings memories of childhood – endless, hot summer days. So easily it happens – one single sighting of a creature and we are sent back to a time decades past. The yellow and black stripes of the cinnabar moth caterpillar on the ragwort and I am a child again upon Holkham beach in Norfolk. We each have our own triggers, our own keys in the natural world that open doors to the past, that lead us so suddenly back to long-lost days.

The storm of last night has subdued some of the array of butterflies that were fluttering about just days ago. A battered ringlet rests upon a blade of meadow grass where last week there were swathes of others flittering about the grasses. The large whites seem to enjoy a higher plane, above the tangled web of the grasses. I watch two bumblebees as they go about their business. One is orange-bottomed, the other is a *terrestris*. I see the individuality of each. One dashes from flower to flower upon the ragwort, while the other meanders rather more gently paced. If we could see each creature for its own self, how differently might we treat the beings of the natural world?

Another ragged ringlet alights upon a brown stem of meadow grass and waits for I know not what, and then flutters off and finds another. And I am distinctly reminded of the pathways we take, pottering from place to place with such apparent sense of purpose where really none exists. I follow the ringlet and many other butterflies start to dry and rise. I bend down and peer into their meadow grass world and begin to see the enormity, the complexity of their universe. Two skippers fly. Twenty yards or so away, another orange-bottomed bumblebee is busy with wondrous industry. From one sunny centre of ragwort flower to another, it gathers pollen. On and off. And I wonder if it might be that same individual bee that I saw some moments back that dashed about with equal endeavour. I peer into these beescapes and find one creature, a smaller bee, which is stuck, clinging to a single ragwort flower. A brush of pollen lies upon its head. It has a pale white bottom. I do not know the species, but I know the feeling. It is tired, resting up after a stormy night. Two hind legs stir reluctantly. Perhaps it has been there since last night, drying through the morning and now it clings as the wind buffets it. And for a little while I really feel for that bee.

When I return to my half-drunk cup of tea, beside the field chair, there is a small green aphid, not five millimetres long, who sits where I would have sat and who I rescue in the nick of time and watch as they crawl upon my hand. Six long legs stride along and then struggle over some strands of my hair. Two antennae are longer than its entire body. The right one is twitching frantically. I watch but my attention is torn by the call of a green woodpecker flying into the oak on the field boundary to the west.

I weave my way along the track that I have forged in recent weeks through head-high thistles that have now been flattened somewhat by last night's storm. The beauty of a common blue butterfly halts my progress. I have to watch it a while. And finally, I force my way out of the field to the other world beyond and think as I step back across the green lane, how easy it is to lose minutes, hours, mornings, within the field. Yet they are hardly moments lost but rather moments of wonder gained, moments when the worries of the everyday that leak away life are gone and the immediate beauty of the world is there, clear and vibrant, before you.

RENATURING

14 July
Most mornings I step from the back door and into the field as early as I can. With a cup of tea in hand, I take a brief wander around this patch of wild, following the tracks forged by rabbits and deer. It is me learning to renature myself.

15 July
Crows caw from the tree tops.

17 July
I stand in the field entrance and marvel at the mauves of the sloes that line the branches of brash blackthorn bushes. I linger by the young blackberries. A mewing cry comes from a late brood. I step closer, peer into the thorns to find the nest before abandoning hope. Along the southern edge of the field runs a bank of raised ground formed from the rubble, the soil, the earth, dug out in the building of an extension to the cottage. Some years back, I scattered some meadow flower seed along this embankment seeking to draw colour and flower and there are now ox-eye daisies and primula in early summer. But the blackthorn grows on. It reaches over and takes the light and the bank shrinks and the meadow flowers seem fewer each year. I am reminded that leaving all to nature does not always mean more variety will grow.

24 July
Over the last few years, the blackthorn in the hedge that marks the green lane has sent suckers up as new growth in the field. If I cut these young blackthorn shrubs down, then I can start to make a wildflower meadow here.

29 July
There is a bare square of dried grass that marks the site of the tent pitched here for much of the summer. When I took the tent

Blackthorn

down two weeks back, it had already started to become part of the field, enmeshed in the meadow grasses which had grown around it. The wind had long since blown the guy ropes free of the tent pegs. The pegs had vanished into the soil. As I lifted a rug from the enclosed space of the tent, there was revealed the immaculate coil of a snake. It lay perfectly still. The shock had me jumping back before reason bit home and told me the creature was harmless – a grass snake. I gazed upon its form a moment – the knit of the scales, the spot of yellow on the collar – then walked away, leaving it to wake in its own good time.

1 August
Last night's rain soaks my boots as I follow the rabbit trail through the field to the northern fence. Yet the sun is up this morning, drying the land. I drink tea and stand and simply am. I listen as

RENATURING

the day unfurls under a perfectly cloudless blue sky. So little seems to stir, as though the day has already settled. A crow calls from a field to the west. A dog barks from the next village to the north. Birds chit. A tractor rolls by on the road south. And the horses behind me stand still as statues. One turns his gaze to mine then turns back. Martins fly above. It is a day of still.

I step through the damp grasses and along the newly re-forged trail through the bramble that leads here to this space beneath the oak boughs. Up in the branches above, a robin flits like a flycatcher, hungry after yesterday's rain.

5 August
With bow saw and loppers, I begin the task of cutting back the blackthorn. Within minutes, I am bloodied and torn. The sun beats down. I lie with one hand holding a blackthorn bush while the other works the saw across the two-inch-wide trunk, trying to sever the stem as close to the ground as possible. It is hot, sweaty work.

7 August
Back at the blackthorn.

I run an extension cable from the kitchen over the green lane and trial the aged electric hedge trimmer. It cuts through the stems of the smaller suckers but whines when up against the wider widths of older growth. I have no chainsaw and soon am back to lying on the earth once more, sawing away at the trunks of the thicker blackthorn bushes.

12 August
My friend Mark Mansfield comes over, driving his Land Rover up into the field with his black Labrador, Bob, sat in the passenger seat. We chat over tea in the kitchen then step out and walk while a steady rain falls. Mark lives over at Kersey, a few miles into

Suffolk. He talks of Arger Fen and the project there of allowing regrowth over a vast area of farmland, and how in just a handful of years ash trees have colonised the land, quickly taking over and turning the fields into thickets.

'You can't even walk through it,' he says.

He tells of a patch of land he looked at some years back with a plan to turn it into woodland. The farmer advised that he'd be spending all his time battling the brambles.

Soon, I will know that fight.

We stand by the twenty-foot deep mass of bramble on the western edge of the field. It has advanced some ten feet in ten years. A foot a year. Now it needs to be halted.

Mark tells me how our friend Yalda, who is also developing a wildflower meadow, has successfully scattered yellow rattle seed. I relay the story of my one unsuccessful try. For whatever reason, I'd not produced the right ground conditions for the rattle seed to grow.

'She's also got orchids,' Mark says.

14 August

There is a foretaste of the end of summer today. Swathes of thistle and ragwort have lost their colour and turned to seed. A breeze blows in from the east. Pigeons tear from the tops of the willows. This little corner of England feels more shrunken today. I step out. The tracks and trails of the footpaths of other creatures guide me as I feel my way. A hornet rises before me, gazing with an open intelligence before veering away. I veer, too, off the rabbit-track pathways into the heart of the field. I stumble on the clumps and bumps of the earth, bend and peer into a brown hole in the ground. I work my way gradually north, touching the dry leaves of the young oaks and remember their feel in spring. Large white butterflies flutter by. A pale moth rises from the meadow grasses and collides with me and flies on. I become merely another item in the field.

RENATURING

The thistledown is sticky to the touch. It collects in clumps like cotton. To halt and pause and delve down into the field here a while is to become a part of the field. An intimacy builds with the creatures. A half-inch-long hoverfly works its ways about the yellow disc of a daisy. A single ladybird sleeps beneath the green blanket of a thistle leaf. A miniature spider rests upon my arm. If I were to simply stop and lie here, so soon, I too would be another creature in the field. Merely that. I would become enmeshed in spider thread and thistledown. I would become another bump within the meadow grasses.

I weave a path to the far fence and halt there in that familiar place. It doesn't feel that this field experiment will ever be complete. I try to imagine, to picture a day when these oaklets are fully grown – 200 years from now – when all this patch of land has turned to trees, as all will in time if left alone to do as it will. For the moment, it seems that if left the field would become entirely populated with thistle. Yet given enough time, the thistle will not cope with the lack of light and space beneath the trees. I close my eyes a moment and wish that I could get just one peek into this place after it has run its course of time and returned to true wild.

16 August

It is a rather grey day in August when I finally get to meet up with Richard Brown. He is the scything man, or at least that's how I have come to think of him.

Richard lives up in Norfolk, a few miles west of King's Lynn. I drive from my home across Suffolk until the land flattens down. The journey north takes me up through lands I do not know – through the King's Forest past Bury St Edmund's and on, beyond signs to Methwold, to Stoke Ferry and Crimplesham, out to dark, flat, fenlands. I pull off the A-road and follow Richard's directions to his house, turning down a track through a wood until it opens to an entrance way. Before me is an aged mansion. I get out and go to a rather grand front door. A stone heraldic shield sits above the doorway. I read the date of 1548.

Richard appears and guides me over to the meadow behind us, to a recently built round house. The roof is partially thatched. I duck down and step inside.

'Tea?' he asks.

There is something rather wonderfully eccentric about the scene. A delightful cornucopia of pots and jars litter the shelves about me. I sit in an aged armchair.

Sweet woodruff, I read on one label.

Settled with a cup of herbal tea, I feel instantly at ease. I want to hear from Richard on his meadow, on scything.

'When you scythe a meadow, you get such a powerful sensory experience,' he explains. 'You *feel* the effect of cutting at different times. You engage with the meadow. It gets ingrained in you. I really didn't expect that at all. That was one of the most amazing things. Revelatory.'

'It's genuinely immersive, isn't it?' I say. 'I've done a bit of scything before, a few years ago in the field. As you say, you're working away and it's kind of meditative, isn't it?'

'Exactly.'

'And also, if you suddenly see a buzzard in the sky above, you can actually hear it as there's no tractor engine rumbling around, or a strimmer. A lot of people use a strimmer to cut down their grasses or meadow, but they make so much noise.'

'They're horrible,' Richard agrees.

'You can see how people get an emotional connection with scything,' I add.

'Absolutely,' he says with a laugh. 'Well I certainly got one.'

He tells me the tale of an old English farmer who had continued to use the older traditions of farming on his smallholding, such as scything, even though many of those methods had largely died out as those around him were using more modern ways. When he had died, he had his scythe placed on the lid of the coffin as it was lowered into the earth.

'There's an expression that says, "I'd no sooner lend my scythe to another man, than my false teeth".'

Richard tells me how in Yorkshire there was a tradition for the farm workers to go to the blacksmith in the spring before the hay harvest to have their scythes individually fitted to make sure they were right before the season.

'Like a dental check-up,' he adds.

Each scythe was made to be worked by an individual farm worker. That was why in the tool collection of the old estates, there would be forks and rakes and all the other implements, but you wouldn't find any scythes. They were like the personal possession of each worker.

'It would be a lifelong connection.'

You find that same tradition all around the world.

I tell Richard of chatting with my friend Selfie the week before in the field, who had worked in orchards for years and who told of the old boys who each kept their own scythes, and kept them so sharp, too.

'That's another reason why you don't lend it to anybody,' adds Richard, 'You don't want to risk them dinging it, or doing something silly with it.'

We laugh and sip woodruff tea.

'There's an English – or Anglo-American – style one, and then a European one,' Richard explains, and reaches for the scythe behind him.

'This is an Austrian one. They come in various forms, the European ones, but they're generally lightweight and very ergonomic. The English/Anglo-American style is much more heavy-duty.'

Some argue that the reason for that is because English grass is heavier, so you need a heavier tool to cut it, whereas there is lighter grass on the continent. While he agrees there might be some truth in that, Richard has his own theory that the real reason may come down to the scale of the farm. Historically, in England, there were larger estates and these would organise the teamwork, with some workers doing the jobs of raking and making the haystacks, while the scything would be done by the most athletic, fittest young men. They would each be given a big, powerful scythe so they could drive through the wheat.

'In areas like the Dales, where there was too much work to do at harvest time, and you needed everyone just to rake and turn the hay on the farm, there were hiring fairs in towns like Skipton,' explains Richard. 'And there would be dedicated scything gangs – often Irish navvies – with huge scythes and muscles, because they did that every day, and they would be hired in to come and cut the grass.'

It was a fabulous slice of historical farming practice in England.

'So that would suit a heavy-duty, high-horsepower scythe,' he adds.

In Europe, and elsewhere where there were smaller farms, there would be a quite different way in which the land was cut, so the needs of the scythe were also different.

'In Romania, where I was recently, each smallholding had a house cow which the family got their daily milk from. There might be other cows on the hillsides, but that cow would need feeding every day. They cut fresh grass. Everyone in the household needed to be able to step out and do some mowing. There's not just one fit young man who does all the cutting.'

'So the scythe needs to be lighter,' I say.

'George Peterken in his book *Meadows* actually goes so far as to say that until the invention of the scythe there was no such thing as a meadow. This particular assemblage of plants only came together with the ability to close a field off and leave it so it wasn't grazed but was allowed to flower and then it could be cut for hay and stored for the winter.'

The meadow such as we would imagine and see today is actually only 2,000 years old. It is a human construct – one forged by the scythe. Of course, the plants always existed in the landscape, but not all together in such blocks, nor in such numbers and concentrations.

It's a wonderful moment of insight. I simply hadn't understood that basic truth about meadows, or their existential connection to the scythe.

'The scythe and the meadow are two sides of the same coin,' Richard says. 'The word meadow comes from the verb "to mow". The meadow is a grass field that you mow, whereas a pasture is one you graze.'

He tells how he uses the scythe as a tool for explaining how ecology and meadow management work side by side and how important the mowing is to sustain the meadow. He explains how he has been a little frustrated by recent movements such as Plantlife's campaign for NoMowMay.

'Not because there's anything wrong with that message about not cutting in May – that's exactly right. But the dear old public always like simple messages. They'll pick up the "No Mow" bit and forget the "May" bit. They'll just think mowing is bad. Just like the idea that cutting trees down is bad. It's too simplistic and, of course, the media like simple messages, too.'

A number of Richard's attendees on his scything courses have been people who have let their lawns grow tall and then they can't work out how to cut it as their lawn mowers won't do it.

'One poor guy, he said, "I persuaded the wife to let the lawn grow tall, and I thought I'd just stick the mower over it at the end of the growing period, and it wouldn't have it. It just stalled all the time."' We both laugh. 'He said, "I had to cut it all with a pair of hand shears. I cut the whole lot. It was dreadful. There's no way I can do that every time. It took so long."'

We laugh again.

'He came on the scythe course and at the end he said, "When I go home with a scythe, that patch won't take an hour."'

Scything is becoming increasingly popular. Richard tells me how there are about 3,000 scythes being sold every year in the UK.

'Really?' I blurt out. 'Isn't that fantastic?'

'It's amazing. From a few hundred to three thousand,' he repeats.

'Isn't that a lovely example of how individuals want to be part of the movement to bring back nature!'

Richard nods. After all, he has been at the heart of this for a long time. He has been working for many years with wildflowers, encouraging others to renature plots and patches of land though his role at Emorsgate Seeds.

'The problem is that we've broken our connection with nature,' Richard says. 'We've got to rebuild that, got to foster that.'

'Absolutely. And there are positive signs.' I'm remembering an article about a recent Chelsea Flower Show entry.

'Someone put in a field of ragwort,' I say. 'And it got a gold award.'

Richard laughs and tells me he has only been to Chelsea once, with a community group that had a sponsored garden which he had helped design and build.

'That year every garden seemed to have a weedy lawn,' he says. 'I thought, "the message is really getting across".'

His concern is that it might only be a fad, that in time the designers might try to force another way.

'There's still a long way to go,' I say, telling Richard of my visit to Anglesey Abbey gardens near Cambridge the day before, with my mum. '. . . There were these huge, perfectly mown lawns.'

'Personally, you'll see with my garden, I quite like the weeds,' he says. 'Though there's more weeds than I would tolerate if I wasn't going out teaching scything! But I know why people have gardens like that – it's a reflection of their need to control their environment. It's a reflection of their personality. For some people, that need to control is so intense, their lawns have to be immaculate. They can't tolerate a single weed.'

We laugh again.

Richard lifts his teacup from its saucer, a little chime of china as the two touch. There's something so deliciously rustic about the entire scene. We're sat there in a wooden round house, in a lovely armchair sipping tea and chatting about old country ways. We've been having such a hoot, the afternoon has just run away with itself.

There's another cry from a buzzard high in the skies above us.

I put my cup down and thank Richard again. We rise from our seats. Time to actually do some scything.

Outside, the day is still grey. There are a series of hayricks, humps of cut hay that look a little like ancient creatures, woolly mammoths perhaps. Against one of these lies my scythe, next to Richard's. He heads off to get a couple of tools in order to adjust the blade on mine so it's more suitable to my frame.

He's shown such generosity to me – inviting me up here to his home, and telling me patiently of the ways of the scythe, both his journey and the history of this remarkable implement.

I stand beneath cloud. The buzzards mew loudly. They must have a nest in the beech tree, I think.

Richard returns. He holds my scythe in his arms and adjusts the bevel that fastens the blade so that it sits at the correct angle.

'The cutting edge is kept slightly above the ground,' he explains. 'The tip of the blade is higher so it will ride over obstacles, rather like a ski.'

There is a swish as the scythe cuts through the grasses and

wildflowers of the meadow. He works away. The sound is oddly pleasing – especially against the backdrop of the thunderous roar of a jet plane above us – a rhythmic, almost mechanical, noise touched every now and then with the metallic ting of the blade catching something more solid in its way.

'I'll let you have a go,' he says. 'I'll just re-sharpen.'

Richard turns to sharpening up the scythe with a whetstone. The edge of the blade sings out with each brush of the stone. There is something in that metallic ring that is so strangely calming. In time, I will get to know it as the sister sound to the swish of the blade through the meadow, the stationary echo forged by hand to the motion of the scythe through the field. For now, I'm happy to simply watch, learn.

Richard steps into position to show me.

'Knees slightly bent,' I note.

'Yup. You don't want to lock your knees. Keep them slightly bent. You're slightly rocking from one side to the other.'

He shows how he starts with the weight on his left foot, then shifts it to his right as his hands bring the scythe across. It would normally be a morning's training to get the essentials. We have a few minutes.

'The key thing is keeping the blade on the ground all the time,' he says. 'Circulating it in an arc. What you want to do is slice along the stems rather than chop it down.'

Richard pauses.

So now it's my turn to have a go. I stand, right foot forward, adjust my feet, then bend my knees slightly before turning the scythe in my hands, trying to ensure the blade swings smoothly and close to the ground. There's a chime as the metal catches the earth. I swing again. And again.

RENATURING

18 August
I step into the field with the scythe over my shoulder.

20 August
August sunshine sneaks between thunderclouds. I step out from the kitchen and down the back of the house to the green lane. I step up the rise into the field and head for the teasels. There are two of them – each six foot high – the cluster of barbed heads bob in the wind. Looking into the inscape of their spiky world, I can see the seeds already starting to shake their way out. I snip off three heads of teasels and catch them in a brown paper bag. They're for my friend Yalda – a reciprocal gift of sorts for that white envelope of bee orchid seeds gathered from her front garden in sleepiest Suffolk. That package still sits on a sideboard in my kitchen. I wonder where the best place in the field is to scatter those precious specks of promise.

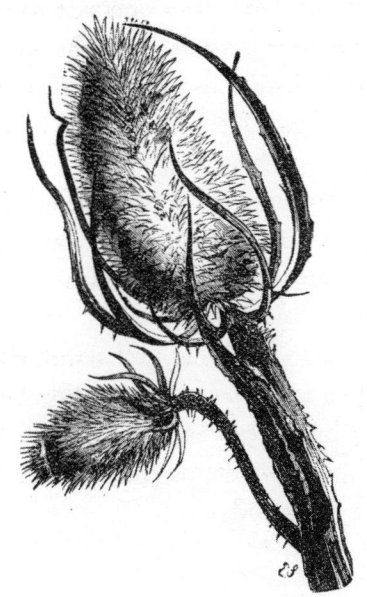

Wild Teasel

5 September
Back from time away in Norfolk, the field looks like a wilderness. I step up the slope and see this patch of scrub with fresh eyes.

7 September
In warm sunshine, I work away at the blackthorn. A small clearing has emerged.

8 September
I come across a line from Zeno – an early Greek philosopher from the fifth century BCE:

> The goal of life is living in agreement with nature.

14 September
My morning ritual now is to step into the field with a cup of tea then walk the deer trail that has become a footpath of sorts through the thick meadow grasses.

17 September
An hour battling with the blackthorn.

I have decided to abandon modern ways – no extension cable and noisy hedge trimmer today – and return to the physical practice of loppers and bow saw. Within minutes, I am glad to have done so. I am so much more a part of the field working like this, my presence so much less intrusive without the noise of the electric machinery. I notice the life about me as I cut and lop – the brush of a breeze against my skin, the call of the rooks, the appearance of the robin beside me peering wide-eyed at the freshly opened mossy ground.

RENATURING

21 September
It is a mast year, for sure. Acorns abound. The ground is gradually cleared of blackthorn. I tear at the moss and the grass, pulling up clumps as though my hands are ruminant mouths – exposing the earth to the air.

22 September
The first day of autumn. A glorious day. Late summer sun lingering on. An Indian summer.
    In the field, I stand by the bank where the yellow rattle grows, clearing the way for more meadow flowers.

26 September
Blackthorn clearing for an hour in the afternoon.
    With practice, I become more swift and efficient – leaning low to the ground and lopping the lesser stems that open the space to allow me to work the bow saw on the inch-thick girths of the older blackthorn growth. I carefully drag away the cut bushes and stack them upon each other, watching for those vicious spines that stick out from the severed shrubs.

30 September
My friend Paul Gwynne pops over around noon with birthday gifts: beers; one of his legendary home-made soups for lunch; and a packet of yellow rattle seed.
    We step out the back door and over the green lane to the field. Berkeley, his dog, leads the way. It's been a while since Paul has seen the field.
    'Wow,' he says and laughs.
    I start to explain how I'm cutting back the secondary hedge of blackthorn. I can see what Paul's already thinking.
    'Should keep me busy,' I say, and laugh too.
    'Is that the black walnut?' he says, pointing over at the tree.

'Yeah.'

Berkeley has headed off to the furthest reaches, smelling deer no doubt. Paul looks puzzled.

'I thought the walnut was over there,' he says, pointing now over to where the cedar towers above all.

Paul's right.

'It was,' I say and laugh again.

The time between seems so much less than it has been.

'I planted it over there when I first got the field, as a memorial to my dad,' I explain. 'At that time, I'd agreed to have horses on the field for grazing, and one of them pulled it up. I walked out one Sunday morning to find the tree in two pieces. A horse had bitten it in half.'

Paul nods as though vaguely recalling the story.

'I bound the two parts together with twine . . . and replanted it there.'

We look over. The black walnut is a glorious arch of branch and leaf reaching thirty feet and more into the sky.

'I didn't think it would make it so I bought that one—' I gesture over to the cedar '—to remember Dad, instead. I thought it was a Lebanese cedar.'

'Oh, yes,' says Paul. 'And how big was it when you bought it?'

I hold my hand at waist height above the ground.

Only later, when fire-side, do I think again of the walnut and remember that the reason I'd wanted a walnut tree to remember Dad was because of one that grew on a neighbouring allotment to his in Cambridge. For a year or so after Dad had died, I kept on his allotment, travelling up every couple of weeks on the train from London, seeing Mum and then pottering about Dad's plot, sitting in his shed when it rained and drinking tea. It was probably all part of the grieving process, though I don't think I would have seen it as that at the time. Then, it seemed a part of him that I could keep going, even though he had gone. I dug the earth, turned the soil. It helped.

One of the friendly fellow allotmenteers – a rather natty-looking woman who must have been in her sixties – would exchange a few words. A walnut tree stood between our two plots on a patch of ground that had been let go. The tree was dead, a skeleton standing as a splendid structure, a bare, long-limbed figure leaning above us. One day, I asked about the tree and the plot next door. The woman explained how the man who had worked that plot had planted the walnut tree himself many, many years before. Some time back, the man had died.

'When he died, the tree died,' she said.

It was strange. I had completely forgotten the story, but talking to Paul earlier in the field had somehow opened some doorway in my mind. I could quite clearly hear those words. Those were the exact words which that woman had said to me over twenty years before.

Her telling me that story of the old man and his walnut tree had been the seed for me buying the black walnut to remember Dad. Not that I had known it was a black walnut at the time, but that is another story entirely.

2 October

It is unseasonably warm and the cottage is awash with ladybirds. It is their movement day. They seem able to sneak in between the seals on the back door. They crawl the walls of the kitchen seeking favoured corners of the windows. They are busy finding their settling-in place for the winter. There must be 200 of them in the porch alone – tucked down in an orange mass of beetle bodies.

I leave them to nestle and escape to the field.

5 October
I step into the field with a scythe over my shoulder. It is a strange feeling to carry such a tool – the obvious resonances to the dark figure of Death are hard to avoid. It is an impressive blade, too – still sharp across the foot and a half of cutting edge. Within a moment or two, I have forgotten about the Grim Reaper. I focus instead on the action of bringing the blade over the ground in a smooth sweeping motion, an inch or so above the soil, neatly slicing the long, wet grasses, the thistles and the tentacles of bramble that have reached far out into the field. When I turn to tackle the bank of bramble, the blade no longer scythes smoothly but rudely snips and slices, crunches into the thicker limbs of the bramble. I slash away. The bare skeletal framework of the bramble bank is exposed gradually as the younger growth is cut back. I know the blade is not designed for such a task and when it jars against a knot of old, tough branches, I wince and in some strange way feel I need to apologise to the scythe for making it strive away at tasks beyond what it was built for.

After a few minutes, I halt and take my worn old jumper off. It is warm work. I stand the scythe beside me and clean the blade carefully between gloved thumb and forefinger. For all that toughest of bramble, it is not damaged.

6 October
Now that the sun has come out, the ivy is alive with wasps and flies of all sizes and occasionally hornets, bumblebees, honeybees and even, I think, the odd ivy bee. These are a recent migrant bee which seems to have moved north with the changing climate, just as the tree bee did a few years ago. I have only seen them in photographs – their body a striped imitation of a wasp's, their head and buff, gingery ruff like that of a honeybee.

I watch this flurry of insects busy feeding on the ivy flowers, the last significant source of nectar and pollen for them this year.

RENATURING

I will not be cutting this hedge in the immediate. It is too cruel. It is no hardship for me to wait a couple of weeks. Yet such a simple act of recognition and then delay may make all the difference to hundreds, thousands of insect lives.

But I am not here merely to gaze upon these creatures. I am here to work. I dip into the shed and seize the scythe.

10 October
I have found that passage in *Anna Karenina* where Levin starts scything a field in a long row of farm workers:

> He heard nothing save the swish of scythes . . . the crescent curve of the cut grass, the grass and flower heads slowly and rhythmically falling before the blade of his scythe.

Once Levin settles into the practice, he muses on the meditative nature of scything:

> The longer Levin mowed, the oftener he felt the moments of unconsciousness in which it seemed not his hands that swung the scythe, but the scythe mowing of itself, a body full of life and consciousness of its own, and as though by magic, without thinking of it, the work turned out regular and well-finished of itself. These were the most blissful moments.[1]

11 October
I step out on a cold, crisp Sunday morning. Somewhere some miles away someone is shooting birds. Here there is quiet and the chit chat of robins. A text pings in from my friend Dave the Bookseller:

> A fine morning in October.

He tells me they are from Dorothy Wordsworth's diary from 11 October 1800.

When I step into the field, my efforts at cutting back the blackthorn are evident. The early morning sunlight has picked out a young oaklet, four, maybe five years old, which is now visible in this newly cleared fragment of land. Beside the oak, a great green clump of ox-eye daisies are now open to the light. I stand and look down through the longer grasses to the emerald leaves tight to the earth. A few feet away in a wetter section, there is a patch of watermint which I will look to nurture. Here the field turns to a boggier corner. There are clumps of sedges and reeds. I look around and start to delineate more discrete sections of the field.

With cutting down this blackthorn, I am starting to actively manage the field, to renature for the first time. Here I will create a section of meadow full of wildflowers by the spring. I have stacked the cut blackthorn shrubs and bushes so as to form a dead hedge, a dividing boundary running some five, six feet high and forty yards along across the field east to west, separating this newly cleared ground from the thicker grasses that make up so much of the field. The stubby one- and two-inch bases of the cut blackthorn trees stick out – have turned from being vertical to horizontal. My hope is that this new spiky thicket will be good nesting habitat for small birds.

We step in and alter the landscape.

Soon there will be a meadow here. It is starting to take shape though there is much more field work to be done before the yellow rattle seed can be scattered.

12 October

In the shed, I find an archaic, aged tool formed of three metal claws fixed on the end of a wooden pole. I have no idea where it came from but it will be great for scarifying the ground.

RENATURING

13 October
Back working at the last of the blackthorn, my bow saw and loppers at my side.

17 October
As I took my daughter Molly to the school bus stop this morning, a van passed with 'Ground Control' written on the side – the title of some gardening/landscaping company. I headed back home singing David Bowie, of course, but musing, too, on stepping into a landscape as a human and directing some sense of 'control' upon it; dictating how things will be, even if doing so from a perspective of genuine desire to improve the lot of the plot for all that live there, seeking to raise the biodiversity of that land.

The central issue of bringing nature back to a plot of land some two acres in size is that a standard method of ground control such as having large herbivorous animals roaming the site is obviously not really feasible. The matter then becomes how to mimic the presence of some ancient lumbering creatures, which is why I have been found in recent days pulling, tearing at the roots of the large grasses, ripping up clumps, sections of the ground cover to expose the bare soil that I will soon sow with my yellow rattle seed.

20 October
Later in the day, I return to the exercise of scarification – the tearing-up of sections of the ground between the bank and the cut blackthorn like an aurochs, I imagine – one of those early ancestors of the cow that stood six foot tall to the shoulder then head and horns on top, whose hooves would have had much the same effect – opening the earth and creating the conditions for seeds to grow.

These vast beasts were sharing the lands of Britain with the hunter-gatherers of the Mesolithic, providing feasts of flesh if they could be killed – if not with arrows and spears, then chased to fall to their death, as at Cheddar Gorge. Their skulls and teeth were found at a burial mound in Irthlingborough, near Peterborough. In a museum in Southend, a couple of years back, I saw the outline of an aurochs drawn on the wall, the scale of these creatures made suddenly real, the shape looming out of the shadows.

I drag my three-clawed tool through the muddy ground, sloshing around inelegantly, sliding about, pulling the handful-sized sods from the teeth and throwing them into the wheelbarrow. Once there is a sizeable mound, I turn to the task of moving the load, pulling the ancient contraption back through the sodden field. I inherited the wheelbarrow from my father. For some reason, I didn't want to let it go when he died. It went to a family friend's shed for many years before I had a place of my own and the space to store it – here. Over the two decades since, it has gradually fallen apart. I have placed various lengths of wood under the great worn hole in the rusted metal cart to keep the thing functional, save its practical value. Today, I pull the wheelbarrow backwards, dragging it through the mud, and finally the whole thing gives way. I am left with only the metal framework – the mere skeleton – in my hands.

24 October
In the last week or so, I have been busy preparing this patch for the yellow rattle – clearing the last of the blackthorn, pawing at the grasses, scarifying the ground. All the time, I have been accompanied by the robin, flitting about in the corner of my eye as I cut back, clear, expose and in the process unveil food from the hedgerow, the earth; coming ever closer to me as I work away.

Now, this scrap of meadow is prepared, or at least as best it will be. I have just scattered the yellow rattle seed, as it is due to rain tomorrow and then pretty much continuously for the next few days. So I sow the seed today and let it settle into the soil with the rain. The yellow rattle seed has to be down to bed in over winter. It needs long months of cold in order to successfully germinate.

Molly came out to help with the sowing of the seed. We stood together in the field and I opened the packet. There felt something solemn in the act as I poured a small heap of the seed into her palm.

We both fell quiet as we stepped about the clearing, heads bowed low, fingers caressing the seed before scattering. There was a sense of the sacred to the action – one born in the ways of the first farmers and inherited through the centuries in some atavistic way hard to explain yet it was there to be felt, that notion of some ancient prayer, some silent incanted pleading that the seeds may rise in the spring.

I saved some of the seeds for my elder daughter Eva who had yet to return from school. When she and I stepped out into the field, that same serious atmosphere fell upon us, the same heavy silence as I poured a pile of seed into Eva's hand, and there was the same sense of mimicking the ways of distant ancestors as we lowered our heads and threw seed into the prepared ground.

So now I stand and survey the scene and reflect. The girls are tucked inside the cottage. It is the gloaming time – that time between night and day when all the creatures of the natural world seem to mark the dying of the light. A couple of pheasants appear, and a squirrel on an oak branch. There is the chit and twitter of robins and wrens.

31 October

A week on and heading out again to go agloaming in the field. Six o'clock and the sun has already set, a very nearly full moon has already risen and the owls are calling.

Yellow Rattle

I step into the sacred space beside the field oak where I have learnt to come and halt. It feels good to spend the time here, merely being, existing by the dark shadow of the oak.

It is a reminder to me to step out at dawn and at dusk, at these liminal times of day, when we can sense some greater knowledge, something more profound about our connection to this earth, to the other beings upon it, when we know better what it is to be alive.

2 November

I stand by the dead hedge of cut blackthorn which seems such a fine place for birds to nest within – the bunched branches look impenetrable to predators. The building of this blackthorn barrier has led to new paths emerging as the deer, the fox and the rabbits are forced to follow fresh tracks. One route runs right by the end of the blackthorn wall, beside a large clump of ox-eye daisies which I have cleared of last year's thick rosebay willowherb stems. It will provide a pocket of colour come spring.

How to improve the lot? That is the thing. For ten years, it was a case of benign neglect, of leaving the field to its own devices. Now, to actively step in and assist in the natural development in order to enhance and extend the biodiversity of the field – to renature – that is the philosophy.

I have opened a path of sorts up through the wide bank of bramble, such that you can sneak around by the earthy bank and the blackthorn hedge to emerge on the other side of the bramble wall in the south-western corner of the field. With endeavour, you can forge your way along the western edge of the field. A chain-link fence marks the boundary, with the water reservoir on the other side. Field-side is a line of oaks, along with the odd ash and some elder trees. It is a fight to get through. Dead brambles and wild roses snag at you. A sodden ditch runs the length of the field edge. It might be man-made or a winterbourne stream. Only in highest summer is the ground there dry. I have never ventured north beyond the last oak.

The colours of the oaks are turning. Some leaves remain green, most now have touches of autumn. We must turn with them. The trees turn inward. That is what we must do too in winter. Some gentle self-reflection through the darker months.

I turn and head in.

3 November
In the meadow patch, I have just thrown down those seeds I have collected from my wanders over the last few weeks – mullein and purple tansy and chicory.

7 November
A stunning sunrise. Once Molly is on the school bus, I step out to sunlight after two days of cloud. It feels like a blessing.

I walk down the lane and wander about, dazed by the light, ending up in the field where I watch the low sun on the oaks; it creates such a warm, honeyed hue to the leaves. My presence startles. There are noises in the undergrowth, movement in the bank of bramble.

A sneeze rises, explodes from me. And as it does, a deer flies from the cover – a muntjac that tears across the field to the unexplored corner in the north-west where the deepest thickets reign. I laugh and follow the path, unhurried. There is a dray up in the roof of an old elder tree. A pale, white underbelly is revealed as a young grey squirrel stretches and plays in the sun. A memory returns from my childhood of a man in a local London park with squirrels in his pockets and handfuls of nuts – a bearded tramp whose face I can almost see now; his worn, black coat hanging down and squirrels running around him as though he were a tree, scampering across his back and about his limbs, and the man's dark eyes and calm smile as he holds out his palms and the squirrels sit upon them and nibble away at the monkey nuts, the discarded casings falling to the ground. He was someone who had slipped away from society, sacrificing the niceties of life, the comforts for the wonder of truly living in nature. Some atavistic urge to step from modern ways had overcome him – to live simply as our distant ancestors did, closer, so much closer to the land, the trees, to other living beings – sharing their environment as an equal.

I know something of that urge, that wild enchantment.

12 November

One of the plants in the bank has particularly caught my attention. As I rake away leaves and scarify the earth, I keep being drawn to these long, elegant stems of greenery. They are fernlike in their structure – a complex mesh of tendril patterning with each twisted curl of the leaf like that of a meandering stream. The long stems are so fragile. I peer in Alice-like. They are see-through and delicate, soft to the touch.

My knowledge of plants is frustrating me. But I know this one.

'Yarrow,' I exclaim to the empty field.

Achillea millefolium.

I brush at the bank in the field with my fingers, clearing light and space for the yarrow. The robin stands in a blackthorn bush a few feet from me and whistles a long, repetitive and rather mournful lament.

I bend in and stare at one strand of yarrow. *Millefolium.* Thousand-leaved. The whole leaf is finely divided into smaller segments, each of which mirrors the overall shape of the leaf. It reminds me of the intricate patterns sculpted into the walls and minarets of the finest-decorated mosques, such as at Granada, or Cordoba – that repeated, interwoven foliate design. For a moment, I am elsewhere, in the dry heat of southern Spain. Then the piercing note of the robin brings me back.

Only yesterday, I heard the legendary naturalist and writer Richard Mabey speaking on the radio of the delight, the wonder of seeing the detail in nature.

I fall into an enchantingly mesmeric state as I return to the task of steadily clearing the ground – lifting leaves, pulling out grasses, teasing, exposing to the light the strands of yarrow, the leaves of the cowslips that otherwise would have been hidden all winter. The robin sings beside me. I work away, bent over the bank of earth and moss and grasses until my bare hands can no longer take the thorns of the thin cords of bramble. I straighten up. The robin falls silent but does not fly. I see how close he is. Perched on a blackthorn branch, he turns his head in that way they do, a bright black eye staring out. For a moment, we both stand still. Then I nod and walk away.

17 November
In the far corner of the meadow, it is a mud bath. I have been busy having a go at hacking out the bunched roots of bramble while imagining my actions as an imitation of the probing of wild boar whose tusks and snouts I have exchanged for mattock and gloved hand. It is dirty, messy work, especially in the dying light and drizzle, sliding about in the mire as the lights in the neighbours' house grow ever brighter in the murky gloam of dusk.

19 November
I am sick of the mud.

21 November
Cowslips are on my mind. A few grow along the north-facing earth bank in the field under the blackthorn hedge that divides the field from the green lane. Cutting down the overhanging branches of blackthorn and bramble, clearing away the dried detritus, pulling up some of the longer grasses that have chocked the bank, will bring in more light so the cowslips can spread and flourish.

Cowslip

RENATURING

23 November
The sun shines. I step out from the workplace indoors to the outdoors, slip into the field and settle at the bank, start to clear and lift with bare fingers – picking the maple and the oak leaves from the surface, from the mosses and patches of bare earth. It is about clearing light and space for favoured plants. I can see newly freed cowslips. I work away with nimble hands, bent over the bank, pulling at the grasses, collecting and throwing handfuls of leaf litter away into the barrow beside me. My fingers become ever more skilled within moments until they seem to be operating independently of my commands, soon taking control of their own movements – speeding at their tasks of selecting and picking unwanted detritus; pulling, plucking at grass stems with thumb and forefinger, gripping around a stubborn stem of bramble, flicking away a worm cast, occasionally feeling the sharp prick of thorn either old or new. My eyes dart and flicker, scanning the surface, picking out the features of the bank, the detail they seek. There the first little leaves of an ox-eye daisy emerging through the mosses. I pull at the grasses that compete, tug and rip them out. Within a minute, the robin is down there beside me – spirit of the bank – flittering and twittering, a presence at the edge of vision, then gone and only a long-drawn note calling from the blackthorn. Then there – a splash of rust, two black beads of eyes staring intently. I say hello. It would be rude not to. Then I look away, back to the bank, work on – lost so soon again in the task of lifting off the fallen fragments of the autumn trees.

What is so remarkable is the effect of this task on me. All thought fades. There is no consideration but for the action of my hands busy over the surface – a close, detailed focus on the ground before me that demands all my attention and in the process shuts down all superfluous thinking, all unnecessary chatter in my mind. I feel as though transformed to some medieval monk – the ancient knowledge of the elders who would work their garden in silence and know that blissful solitude of being immersed in nature. We know such simple practices are so healthy for our wellbeing, and yet it can be so hard to find the time in the busy background of life to take ourselves away from the other demands of the day. I think how I could happily keep doing this for hours.

Another young primrose, hidden. I pull back the mossy mosaic around it, open the plant to the light. I pick off the brown frames of leaves surrounding with swift, practised hands. The moments pass. I keep working away at this simple task of carefully tending the garden and know that sense of delightful detachment that Buddhists call *dhyana* – a meditative state whereby the normal rattle and chatter of thoughts falls silent, and perfect equanimity reigns instead. I smile. The robin stands on a branch at eye level three feet away. He tilts his head as I raise mine.

24 November
Back in the bramble – slashing away with the scythe then turning to the mattock for the roots. Within minutes, I am slipping and sliding around as though on an ice rink.

25 November
Late morning, I am head down, staring at the ground and suddenly struck with the stunning beauty of the young stem of a yarrow plant: that same delicate fretwork structure, yet in miniature. I carefully tease away the moss around, open it to the light.

I hear the sound of the robin before I see him; not his song but the brush of his feet on the branches beside me.

'God is in the detail,' I say out loud. I am still staring at the perfection of the inscape of that inch-long arrow of yarrow.

The high-pitched call of birds above turns my head. The sound rings from the blackthorn. At first I take it as the fluted chat of long-tailed tits. Yet something tells me it is not them. The pitch is even higher. I see the flit of movement, a tiny bird that flutters in the hedgerow, then another close by, a dash of red distinct upon its crown. Two. Their calls tinkle like angel bells.

'Oh,' I say, the sound falling from me, spilling out.

I have never seen or heard them here in the field before. Goldcrest.

For a few seconds more, I see them, hear their presence, and then they are gone and the silence they leave is one full of sadness. It is a brief moment of mourning, the air bereft.

Then the robin is back and stood before me and I am all vision. I see him and smile. He turns a head and then he flies.

2 December

When the sun rises today, it does so clear and wide and full of hope for the future, bringing clarity and light. There was another cold night. The cut blackthorn stacked in the field is pale with frost.

The oak beside them gets ever more beautiful, with its bronze leaves that refuse to drop. Different oak trees' leaves turn to such a variety of shades and colours. Some stand bare while others wear copper cloaks. 'Phenotypic variance' is the scientific term – individual genetic differences within the species – which serves to remind us that each and every tree is an individual living being, each one distinct from the other, just as we humans are.

Three pheasants fly. A wren chatters in the copse.

As the sun rose, it was mirrored by the full moon sat in the north-west of the sky. It feels a profound time of year – stepping towards the turning time, the delicate balancing of the light and the dark, and the sun and the moon, and the day and the night.

4 December

Snow.

Snow on the grass and on the roofs of the houses. Snow on the fields and on the cars and on the trees. Snow everywhere.

Unexpected, it has fallen in the night and we wake to a white-out. The spectacle is not lost on me. It is a thrill and a joy that rises on seeing the white blanket beyond the window. Soon, I am out. In the field, there are inch-thick layers of snow on the boughs of the young oaks; bowing crooked the frame of the blackberry bush. The sky is a uniform grey. I walk and hear only the crump of the snow.

It is a school day. I will have to clear the car to take Molly to the bus stop. I head in, have a quick breakfast, then am out again sweeping armfuls of snow off the windscreen, the bonnet. Already it is turning to slush, a great thaw underway. The green lane is dripping. In the neighbouring field is the figure of Tony next door, out with the greyhounds.

'Wasn't expecting this, eh?' I call.

He laughs and tells of roads blocked, trains cancelled. It doesn't take much to bring the country to a standstill. The dogs run around like excited children. I leave him to walk them round his perfectly white field.

By the blackthorn dead hedge, I greet the robin, feeling for his plight. I wish I had a handful of mealworms to offer. Yet the world is soon melting. I step where pathways once were. The arrowhead footprints of deer in the snow halt my progress. I am forced to stare. When I look up, the bronze leaves on an oak tree are startling – stark and vivid against the white backdrop of the field.

5 December

The snow has gone – as if it never came.

8 December

Back to tackling the bramble along the western edge of the field. I dress as though to go into combat.

RENATURING

10 December

Working away with loppers at the bramble bank, I am halted by the sudden presence of a family of long-tailed tits. I stop and watch – awed by their industry, their joy as they dash about the oak branches above. This is one benefit of working without power tools – seeing the silhouettes of other beings as they appear before me, being present to the life of the field even while chopping and lopping.

11 December

The bramble bank is huge. It runs the entire length of the western edge of the field – some 300 yards or so. Once the blackthorn had been cut down and the first meadow patch sown with yellow rattle, the plan was to hack the bramble back so as to make another, longer meadow that would stretch the extent of the field. Already that feels an endless task.

17 December

A day of sunshine. In the field I chop away with shears at bramble, releasing the limbs of a young oak from the barbed tentacles which have held it down for years. Though winter is yet to come, there is that stark, cold clarity to the world.

In the western oaks, there are fieldfares. Half a dozen or so, perched nervous in the upper branches. I watch a clutch of blue tits flit through – picking, pecking, snacking as they go. The fieldfares fly. A great spotted woodpecker sits halfway up an oak, framed in profile against the blue sky.

There was shooting yesterday. Guns blazing. I saw a pheasant flying over the house from the south in the afternoon, fleeing. The field is a refuge to so many of the creatures around. I watch one male pheasant now as he steps delicately over the ground being cleared for meadow flowers. He strolls and picks and pecks with an ungainly movement of the head, white collar jerking. He seems to have found a home here.

I drag an old pallet – scrounged from the house over the road many years back – out of the bramble and use its half-rotten frame for drying logs in preparation for the next bonfire. There is the trunk of a laburnum tree which I cut and dug out of the garden ten or more years ago. I pause a moment and see Eva, as a toddler, stood underneath the tree in an umbrella of yellow flowers. A decade on, the weight of the trunk is impressive. I cut a section off with a bow saw and feel the strength, the density of the wood, even after all this time left out in the field.

20 December
Working at the bramble bank for an hour in the half-light of the early afternoon.

21 December
On this shortest of days, I stand in the field as the sun rises over the copse of oak and goat willow. Later, I will watch the sun set, bear full witness to the winter solstice. Tomorrow the light will start to return.

23 December
The dawn seeps in through grey cloud. In the field, I stand under the evergreen pines of the cedar. Two goldcrests dash about the branches above me. There are half a dozen blue tits darting about with them. I stand in awed silence as gentle chimes fall.

1 January
I lurk in the field at dusk, tucked away beneath the largest of the oaks by the western edge, and simply stand, listening to the chitter of the birds preparing for the dark of night. Wrens tic.

RENATURING

5 January
Before I have yet awoken, I step across the green lane and into the field and stand there in silence, broken only by birdsong. I am enchanted – held in this moment, in this august space. Where once the blackthorn grew, now there is an empty patch of cleared ground, some twenty yards across and forty, fifty yards long.

Oaks rise, half a dozen or so, none more than a few years old, a couple yet to see their first winter. A male pheasant sits at the very top of the cedar tree. No fox will take him there. Nor gun either. I try not to scare him by my presence in this still time of day. It is no good – he flees.

7 January
Frost covers the field. The delicate features of each head of thistle are so clearly visible, neatly framed in their frozen form. The yellow rattle will be enjoying this spell of cold. Some stirring within each single circle of seed tells of a meadow in spring, a field of nodding sunlight flowers. The future is already here.

14 January
I step out into the field soon after sunrise. There is no sun. A foggy mist hangs over the frost. From the uppermost branch of a maple comes the weaving, endless song of a song thrush.

18 January
The temperature reached -5 degrees last night. There are drops of frozen water like those that hang from an old man's nose in winter. And as I stand at the edge of the field, the robin who I normally see over on the bank flies right over to me, to land on a branch of the oak only two feet from my face and for a few seconds bobs and stares and then flees. I feel blessed.

It is cold – the second hard frost in a week. I head in.

22 January
The day is dreich. There is no other word for it. From dawn, there is a greyness that seeps into all. It seems it will be a day to endure – one of those days when it will rain more than intermittently until it feels endless, and when sunlight will never really edge out the darkness.

Dreich is a word that is long obsolete, according to my *Shorter Oxford English Dictionary*. One exclusively of Scotland and the north, corresponding to the Old Norse *drjugr*: enduring, lasting.

28 January
It is a wet day, rain already starting to form puddles on the surface of the saturated soil. By the bonfire pit, I find a dead muntjac fawn. The creature lies on its side, the fur plastered down, the eyes already gone to the crows. For a moment, I stand over the body and wonder if I should move it. The rain falls. Lying there in a sodden grave seems no just end for any being, let alone a young fawn. I head off to the shed to find something suitable with which to lift the body, and return with an old length of wood. It will do.

Once the body is on the board, I feel better – as though I have marked in some small way the life of this being. I have already decided the best policy is to leave the creature to the elements – excarnation, as it is known. In prehistoric times, it was a common way of dealing with the dead. The body would be ritually placed on a raised platform of some kind and gradually the flesh would be eaten by the birds. So I have lifted the soaking body onto the board with due reverence and then raised the board up and away from the earth, holding the make-do bier at arm's length and walking slowly, solemnly, over to the stack of cut blackthorn. I place the board and its body on the top, some five feet from the ground and open to the airs above, then back away.

RENATURING

1 February
Cutting the bramble bank back again – armed today with modern weapons. I hold an electric hedge trimmer tied to 100 metres of cable running all the way to the cottage. My armour is on – protection for face, eyes, hands, arms and body as I wade into the morass of aged thorn. My plan is to cut the old growth down at the ground, drag it away to be piled high and burnt – then return in due course with the mattock to tackle the roots in rather drier weather.

3 February
Two hours in a whirring maelstrom of noise. My arms and my head ache but a long barrow of brown bramble has now formed in the field.

4 February
Welcome sunshine. I seize the scythe this morning. Yet much of the old bramble is so tough that I am forced to turn to the ugly roar of the hedge trimmer instead.

9 February
A few days away from the bramble. White finger from holding the monstrous trimming machine means I cannot close my hands. I wonder if an aged sickle I have found in the shed will help.

10 February
The process of buying wildflower seed began in early January. I ordered the seed catalogue from Emorsgate, and when it arrived I flicked through the pages of colourful meadows rather as though wandering through some Edenic landscape, imagining the self-same swathes of wildflower in the field. At that time my plan was to buy enough seed to forge a wildflower meadow from the area where I had cut down the blackthorn and laid it to form a

natural barrier. I had estimated the section to be something like 30 metres by 10 metres – around 300 square metres. According to the guide, the rate of seed-sowing should be four grams per square metre. In the end, I ordered a kilo of meadow mixture for clay soils, which cost £56. That would mean sowing at near enough the recommended rate.

I also bought 200 grams of standard cornfield mixture. Though the patches of wildflower meadow I was looking to create were to be full of perennial plants and grasses, the reality was that I would need more than a single growing season to achieve this. A cornfield mixture would only consist of annual flowers, yet would serve to offer some welcome colour over the first year. The catalogue explained how cornfield seed can act 'as a "nurse" for a meadow mixture'. I loved that sense of nursing a new wildflower meadow through its first year or two, until it had gained more maturity, better plant health. The cornfield mixture consisted of corncockle, cornflower, corn chamomile, corn marigold and, of course, common poppy. With this supplement to the meadow mixture I could hope for a vibrant show of wildflowers in the summer.

Paul had already spoken to me some time before about the use of seed plugs to provide a more secure way of propping up the growth of the wildflower meadow. By seeding various wildflowers by hand, in seed trays, then transplanting them once safely established, there was a greater sense of control over the process. This meant a better chance of success. So with the patient desire of a child in a sweetshop, I pored over the list of native wildflower species available. Each wildflower was displayed in a tiny, centimetre-square colour photograph. The scientific name of the flower was given first in the form of the Latin binomial, then the common name and the 'place of the wild origin' of the seed in Britain. Some names I knew. Many others I didn't. The names intrigued and enticed: ploughman's spikenard (*Inula conyzae*), corky-fruited water-dropwort (*Oenanthe pimpinelloides*) and navelwort (*Umbilicus rupestris*). Some names took me away to the furthest shores of the archipelago: thrift (*Armeria maritima ssp maritima*) to the last vestige of land on the north of the island of Lewis in the Outer Hebrides – the Butt – and a remembrance of lying on a sunny, windswept day beside clumps of the pink flowers, looking out to the Atlantic seas beyond; sea-holly (*Eryngium maritimum*) to the sandy dunes of the North Norfolk coast and endless summer holidays.

In the end, I picked out a handful of wildflowers I most wanted to see in the field and that I had a realistic chance of raising: the purples and blues of corncockle and cornflower; then a punt on tree mallow, devil's-bit scabious and great mullein. With luck, three months on, they would be reaching up from their plugs in the seed trays.

12 February

In the south-western corner, I have been edging into the bramble to reach a special feature of the field that has got terribly overgrown. I began soon after ordering the seeds, simply chopping away by hand with a pair of shears, snipping at the barbed wall and steadily working my way towards the wild service tree. The ground is permanently saturated. I have been slipping about in wellington boots, wrapped up in an old pea coat that was quickly splattered with mud. Yet soon, the tree was freed.

I had planted the tree ten years ago, probably more. Paul and I had talked in the pub one night of putting one in the field. The wild service tree is an indicator species for ancient woodland.

'That would fool some future arboriculturist,' he'd joked.

It had been around the time I had started to not mow the far section of the field and the first jay-sown oaks had begun to appear.

Some months after that chat in the pub, Paul had presented

me with a young wild service tree for my birthday. The problem was that where I had planted it the ground was especially wet, so while the tree had flourished and grown well, it had slumped over. The main trunk was at an angle of sixty degrees or so to the ground.

'Looks like the tree is starting to adjust itself,' said Paul when he came over recently.

He was holding what had once been a side branch but which now, due to the lean of the tree, pointed skywards. It had grown substantially bigger, as though becoming the main branch.

'In time, this section could well die off,' he said, indicating what had once been the top of the tree.

It was a good reminder of how each individual tree adapts to the environment and circumstances that it finds itself in.

Along with the wild service tree, there are two young oaks nearby that need liberating. The bramble has done a fine job of

Wild Service Tree

acting as a nursery to these saplings, protecting them from being gnawed by rabbits and deer, and the oaks have managed to keep their heads above the rising surge over the years. I chop away. It is bruising work – the repetitive snapping of the shears, the scratching of the bramble. Then I realise I could run two fifty-metre extension cables from out of a window in the cottage, across the green lane and into the field, allowing me to turn my old hedge trimmer to the task. The machine tears through the ancient barbed stems. My bramble-cutting practice improves. I learn that by crouching low, I can sever the spiky shoots of bramble close to the ground and then hack through the remaining mesh. Then the severed stems can be gathered into huge barbed bails which burn with a delightful fizz and crackle in wild flares of flame.

Throughout the freezing, wet days of January and early February I've worked away at the bramble. My hands are clawed from gripping the machine. My arms ache from the weight of the thing. But the barrier of bramble has gradually receded. I've realised that once the cut bramble has been raked away, what is left is bare earth. Unlike the cleared blackthorn areas, which I'd had to scarify so vigorously before sowing them with yellow rattle, these patches of bramble clearance are already halfway prepared for receiving wildflower seed. I will create a wildflower meadow right the way along the western edge of the field. And another smaller section on the eastern side. The vision of poppies and cornflowers swaying in the summer breeze is intoxicating.

As I've hacked away, clearing more bramble mesh, working into the early twilight of the afternoons, my outfit has grown more grotesque – layers of old clothes, gloves, coat and hat caked in mud. The goggles I wear to protect against flying debris soon steam up, leaving me half-blinded, slashing away with the trimmer in the dusk light. I've become increasingly conscious of the young oaks growing up within the protective embrace of the bramble. A couple of them have taken blows, but even in the midst of the noise and semi-darkness, I can sense the odd occasion when the blade strikes a young oak instead of bramble. Each incident is followed by swearing and apologies. Fortunately, none of the freed oaks are badly damaged.

Rather than cart the cut bramble across the field to the bonfire patch, I decide to gather it *in situ* – creating a great sinuous heap on the boundary where the bare, freshly cleared earth meets the grass. With huge delight, I realise I've been operating my own slash and burn policy – working the land with an ancient attitude, renaturing the field with a Mesolithic mindset.

Over the worst weeks of winter, I've had a reason to step out into the field and a desire to work away at my task. The vision of a whole length of the field converted to meadowflowers has driven me on. I've bought a second electrical extension cord so as to reach towards the north-western edge of the bramble. Three towering, six-foot-high mounds of bramble mesh are growing skywards, like pyres prepared to welcome the dead.

When snow fell, Molly and I built a snow owl that sat upon the garden table, with wings of ice from the pond. Eva and I headed out over the lanes. We marvelled at the natural ice sculptures which lined the roadside in places where the odd passing car had splashed water onto already frozen branches. The subsequent drip produced a frieze of icicles. We walked over to the burnt oak on Hampers Lane, met our neighbour Betty out walking Nell on the white fields, then turned for home. It was bitter out on the footpath over the field beyond the church. Eva was shivering.

'I can't feel my hands,' she said.

15 February
At the full extent of my reach with electric cable – the end buttress of the bramble will have to remain in the far north-west corner of the field. Anyway, I am done with these thorns. My wrists are scarred with scratches and tears.

RENATURING

20 February
The days have been dry – three great stacks of bramble sit perfectly ready to be lit, as though awaiting some ancient ceremonial event.

21 February
Burning day.

I begin with the brown line of cut bramble piled three feet high that runs some twenty yards, curving along the boundary of grass and earth like some monstrous snake. My plan is simple: light one end and watch the flames rise and creep along the length of the beast's body. There is a genuine thrill as the moment approaches – the adrenaline fuel of the pyromaniac taking hold. I tuck a couple of scrunched sheets of newspaper into the barbed mesh at one end, click my lighter and step back as the flame transfers from paper to bramble.

The effect is more instant than I'd anticipated. Within seconds, the fire takes hold, flickering at first, then rising keenly, the burning fingers of yellow and orange soon growing fiercely until in less than a minute they have reached greedily up, becoming taller than me, and through which I can see the perfect blue sky distorted by heat haze. There is a crackle that grows in intensity as the flames rise, and every few seconds the bang of an explosion bursts from the conflagration. I watch, enthralled. The flames grow. As the fire burns, so the brown creature that had lain still upon the ground is transformed into something alive. The beast dances with ever brighter shades of gold, copper, amber, marmalade and Mars-orange. As the flames burn, so the bramble turns a ghostly white, and as the fire edges voraciously along the body a pale tail forms, with smaller flames flickering in the embers.

The fire is mesmerising. I follow the flames as they burn through the length of that line of old bramble, spitting and cackling as it goes. There is a sense of some ancient notion of what fire means – the effect it has on the human mind.

In his book *Burning Planet*, Andrew C. Scott distinguishes between wildfire and anthropogenic fire. Many consider it was *Homo erectus* who was the first hominin to hold sway over fire around 1.9 million years ago. As they stepped out of Africa and started to spread into Eurasia, it was the capacity to control fire that allowed these early humans to survive in the colder climates. Yet fire offered so much more than mere warmth. Fire also offered protection from large predators, allowed food and tool preparation and provided the collective focus for social interaction, creating conditions favourable to group bonding and the formation of language. Cooking food allowed for easier digestion. Smoke helped ward off insects. Life was simply so much better for those who controlled fire.[2]

The other valuable impact of burning was on the environment. Fire stimulated the regeneration of plants and that in turn attracted herds of herbivores to the sites, which could be hunted.

22 February

A week or so ago, Paul popped over, dropping off a couple of books. It had been well below freezing the night before.

I showed him the progress in the field before we went for a walk down to Two Oak Hill.

'You've cleared a hell of a lot since I was last here,' he said.

'It's kept me busy,' I joked.

It had been a long, hard winter.

We walked over the cleared ground. It was frozen. Paul nudged at the exposed head of one of the bramble roots.

'You going to dig these up?'

I nodded and smiled. I kicked another one beside me.

'As best I can,' I said.

Paul looked down the section of exposed earth.

'If you don't it'll soon look like that again,' he said and turned his head to the bank of bramble still loosely coated in snow.

I knew he was right.

RENATURING

Some months back, I'd spent a couple of hours inelegantly hacking away with a mattock at the ground. That was in mud. I had slid about in the mire, my boots unable to get a foothold, the handle of the mattock soon too muddy and slippery to hold. It was hopeless work. I tried clearing the wet earth off the handle with tufts of grass but within a few more slaps into the earth, it was covered again. I had returned to cutting back the bramble above ground instead. Now, the ground was frozen. You couldn't even break the surface – even with a mattock. It was going to be a nightmare.

I had already made the decision not to use any chemicals on the field. There would be no weed killer. Chris Gibson had suggested that if I wanted I could put a touch of RoundUp on the exposed bramble roots. I had weighed it up. It just didn't sit right, bringing poison into the field ecosystem. I would simply have to dig up the bramble roots by hand.

There was nothing else for it.

25 February
The bramble pyres rise higher.

28 February
Burning the pyres.

My thinking was that the burning of the ground should also help loosen the grip of the bramble roots that lurked beneath the surface. What it certainly does is help dispose of the knuckles of knurled roots that I had already dug up. I'd spread them across the tops of the other heaps of bramble to dry. Winter winds and sunshine did the job. There are now three great mounds each taller than me ready to burn, and one even larger pile that I've created on the site of where I intend to place a bell tent. Burning is also good for flattening out bumpy ground.

These pyres throw flames wildly high into the cold, blue sky.

I light each one in turn. Each time that same primitive thrill grows within me as the fire takes hold. The ancient past may be many thousands of years away, yet in such simple moments there is a reaching-out into prehistory, a feeling for the hand that lit a distant fire to burn brushwood to open out a glade, with the knowledge that the deer would come and feed on the fresh growth.

2 March
Rain.

In the far western corner of the patch I have now started to call the meadow, mud is now the mainstay. I have begun hacking at the bramble roots there with a mattock, then pulling at the knotty clumps. In dank drizzle, my wellies slide about on the surface as I chop at the earth, making even more of a quagmire of the ground around me until I retire with a wheelbarrow laden with muddy tendrils of tubers torn from the soils, my gloves sodden and bespattered like the rest of me.

6 March
On a grey day, Eva and I put up the bell tent on the cleared ground where the largest pyre burnt. It is a patch of welcome white. We laugh together once it is raised.

9 March
I stand by a sunlit young oak, whose bronze leaves have stayed throughout everything winter threw at us. Surely this is a sacred tree whose susurrations I really should listen to.

10 March
I stand in sunshine in boggy mire, hacking away at the ground with a mattock whose handle is caked in mud and which slides from my gloved grip. I stare at the ground as I work, at the sods of earth, digging out the bramble roots – the knuckles that lie

beneath the surface. I hack and slide, pulling at the earth. Their reluctance to leave the darkness leads to a tug of war, a battle against the physical ache in me until finally each otherworldly creature is untimely ripped from beyond.

They surface as though crazed and with some tentacles still clinging to the dirt. I shake each one violently, thump them against the metal head of the mattock to clear some of the soil and throw them into the new wheelbarrow, where they turn into a pile of strange, giant, underground spiders wrenched into the light.

12 March
An hour with mattock in hands – clearing the ground of ever more dark clumps of bramble roots. Gradually, this ground is opened up, free from the greatest knots, though I know deep within there still lurk others.

16 March
Seed-sowing.

First, I began by dividing the area to be cleared of bramble into sections. The wildflower seed that I had already bought was to turn the area cleared of blackthorn to meadow. Yet I'd soon realised it would be a nightmare to clear it of grass and moss, and dig out the blackthorn roots. And I had sown it with yellow rattle seed back in October. The area freshly cleared of bramble was at least double the size.

By separating it into sections, I could more easily distribute the wildflower seed. It also made the task of digging up the bramble roots seem rather more manageable. I collected an armful of bamboo sticks from the shed and stepped back into the field as a land surveyor. The long strip on the western edge I divided into six sections, measuring each roughly by strides and marking them off with the bamboo sticks. The seventh was the cleared area on the eastern edge running some twenty metres long towards the black walnut tree.

I went back to my study and drew a map of the field with its newly founded wildflower meadow sections, each with their approximate dimensions recorded for weighing out the seed. There was a total of some 400 square metres.

I sat back from my desk.

All I had to do was dig out the remaining bramble roots, scarify the earth and scatter the seeds.

On Native vs Non-Native Species

All across the globe non-native species have been brought to places beyond their natural range by human interventions of some sort or another, whether deliberate or accidental. In Britain alone, there are some 3,000 non-native species. Native ecosystems shift and alter over time. That is healthy. Yet certain kinds of non-native plants and animals that are introduced can unbalance this natural equilibrium. They are called *invasive* non-native species. They are those plants and creatures that once introduced into new worlds are freed from the natural controls of their native ecosystems and so spread and expand rapidly, with devastating consequences for species native to those lands.

A good illustration is tumbleweed – those dry plant balls you see blowing across the desert in all good Western movies. However, tumbleweed (*Salsola tragus*) is not native to North America, but to Europe and Central Asia. Also known as Russian thistle, wind witch and Russian cactus, the plant is believed to have been introduced to South Dakota by accident in the 1870s via seed in a sack of flaxseed. Highly invasive, tumbleweed was soon rolling across the United States and, in time, becoming an icon of the Old West. Tumbleweed loved the freshly ploughed lands of the Great Plains, soon establishing itself across the continent as pioneering farmers cleared the prairie grasses and so provided a perfect environment for the newly arrived migrant. The problem with tumbleweed is that when it ages it turns to a vicious, spiky plant that will harm grazing animals and infest fields, making crops unviable.

There are numerous examples of invasive non-native species and the damage they cause, both economic and ecological, as they thrive in their new ranges, predating on native species, bringing in fresh diseases, affecting genetic diversity through hybridisation and out-competing local wildlife for resources. The issue is massive and global. The United States Environmental Protection Agency estimates the cost of invasive non-native species at $138 billion per year. Over half the extinctions of the 135 species of birds lost in the last 500 years are due to invasive non-native species. After habitat loss, the introduction of invasive non-native species is the single biggest threat to biodiversity across the world.

But the truth is that most non-native species are not invasive. Most are harmless. Most do not disturb the ecosystems or the locals in their new landscapes. Many add welcome variety and colour. The monkeyflower (*Mimulus guttatus*) growing in the tiny pond in my garden, for example, produces a wonderful spray of bright-yellow flowers in spring – each with something of a monkey-like face. But monkeyflower is a non-native. It was first introduced to Britain in 1824 from North America. So the question is, should I dig it up from my pond and replace it with something native? When I asked my friend and go-to naturalist guru Chris Gibson, his reply was along the lines that if insects and pollinators were happily visiting the flowers, then I should leave it be. Monkeyflower has been in the country for 200 years and hasn't acted as a particularly invasive bully yet. Certainly there are other, recent invasive species that are more of a worry. The Asian hornet is one such example. Introduced by accident into France in 2004, these insects predate on bees and have been causing havoc to populations of European honeybees, along with other bee species, both solitary and colonial. UK government agencies are actively working to eradicate any sightings, and environmental organisations such as the RSPB are encouraging members of the public to report any Asian hornets they see.

But as I've said, most non-native species won't do anything damaging or deleterious. In Britain, think of the little owl (*Athene noctua*) – introduced in the 1870s and showing no signs of doing anything more menacing than bobbing its head up and down and staring intently when threatened. Less than eight inches in length, these cute balls of fluff are certainly not a concern. Contrast them with the grey squirrel (*Sciurus carolinensis*) which was also brought to Britain from North America around the same time. These furry visitors are highly invasive. They strip trees of bark to reach the soft inner lining, causing immense damage and they have devastated native red squirrel populations by spreading squirrel pox (which they are apparently immune to). The key is to know which non-native species are likely to be a problem and controlling their spread before they become established – just as is currently happening in Britain with the Asian hornet.[3]

18 March
On the kitchen scales, I weigh out 256 grams of meadow mix – following Emorsgate's guide of four grams per square metre – and pour the brown jumble of seeds into a large blue mixing bowl. Then I add 100 grams of the cornfield mix with its smattering of larger black seeds. I stare at the mixture. It looks rather like an unappealing muesli.

The first section of wildflower meadow is prepared for sowing. I have slithered about for the last week or so, pulling earth-bound aliens from their muddy homes and with the feeling of time ticking away have finally declared the patch ready for sowing.

Now, it is with an oddly sacred sense that I stand with bowl in hand and start to scatter handfuls of seed over the plot of earth before me. My being focuses on the task and there is a stillness, a calm, as I step over the ground, concentrating on ensuring the seed is evenly cast. I cannot help but strive to feel something of what the earliest farmers must have felt as they carried out the same action of sowing seeds on the very first fields. There is a quiet intensity to this simple act. I feel it as my bare fingers tuck into the warmth of the mixture in the bowl, then in the motion of my arm and the release into the air, the shower of seed.

I sow the land.

There is a touch of the Neolithic to the day.

Within this still silence, I gently walk over the ground to ensure the seed is touching the soil, walking methodically over each inch of the patch in a meditative and unrushed manner. I stare down at my feet, feel the rush of attention on this task of ensuring the seed is bedded in and sense again the ancient, ritual nature of my actions.

19 March

It was back in January that the apple tree incident occurred. For some weeks, there had been bites taken out of the bark of some of the outermost branches. The pile of pruning that lay some yards away also showed signs of having been stripped by either deer or rabbits, or both.

Then one Sunday I had gone out to the field and seen that the bark of a section the main trunk of the apple tree had been gnawed away entirely overnight.

Later that day there was a knock at the front door, which was unusual. I opened it to find Tony from next door, wrapped up in his layers and hat.

'Have you seen your apple tree?' he asked.

'Yeah,' I said.

Tony seemed more shocked than me.

'Overnight,' he said.

'I know,' I agreed. 'Not sure what to do about it.'

He opened both hands and shrugged.

'Not a lot you can do.'

I had a look online. The tree had been ring-barked. Badly ring-barked. The term, I discovered, was 'girdled'. My apple tree had been girdled. If the damage had been severe enough to sever the flow of the sap in spring, it would certainly die. I put on my mud-splattered pea coat and went back to the field to check again. There was a section at least six inches long where the bark had been entirely stripped off from all around the trunk. It was worse that I'd thought it was. Properly girdled.

The only possible hope was to wait until the very start of spring and try something called bridge grafting. So I waited. And in the meantime, I did what I wished I'd done on first seeing those bites to the outer branches – rigged up a circle of chicken wire around the tree.

Now it is March. I return to the task of trying to save the apple tree. In the interim there has been the snow of February. Some of the outermost branches still in reach show signs of being freshly nibbled. I check the track marks: rabbits, not deer.

What is odd is that there are *two* apple trees – both planted some fifteen years ago and growing side by side – and one has not been touched by the rabbits. It is a different variety – a pink. Clearly, they prefer the taste of the other.

I search the term 'bridge grafting' once again and watch a video of a Canadian forester repairing a 'girdled, deer-damaged' pine tree that was in a far worse state than my apple tree, with a section of bark some two feet long having been stripped off. Perhaps there is still hope.

It is a crisp Sunday morning. I head out to the apple tree armed with a variety of tools and settle by the tree, clearing away the chicken wire and pondering the task. The notion is to provide a mechanism for water and minerals to pass across the ringed section of trunk. I look up to the various fresh stems shooting up to the blue sky. I'd left them when pruning back in November. Now they will be vital. I cut one and hold the three-foot shoot against the stripped trunk. I cut each end of the stem into a V-shape then lie on the ground by the apple tree. Then I carefully slip one end of the stem under the remaining bark above the tear, having trimmed the bark a little to ensure a crisp edge. With one end in, I bend the shoot and tuck the other end into the bark below the exposed section. There is a bow to the stem but it sits securely enough.

The key is to ensure the stem is inserted the correct way up to allow the flow of xylem and phloem to go from the earth up into the tree. I stand up and choose the next water shoot to cut from the top of the tree, then lie back down and get to work turning it into a second bridge, falling silent, hardly even breathing as I concentrate on the operation in hand. By the fourth bridge, I can feel my practice improving. It is remarkable how much more expert each bridging is – a sharper eye, an understanding of the task, a confidence of cut. I carry out a fifth bridge then stuff patches of moss soaked in rainwater around the spliced grafts to keep them damp.

I stand back again. There is a sense of a modern art installation to the structure. I feel sure that keeping the bridging joints both secure and moist is going to be crucial, so wrap strips from an old towel round top and bottom with grey duct-tape to keep it all in place. Then I put the chicken wire back.

There is something distinctly uplifting in having undertaken the procedure. I tidy up the tools with a light heart.

Even if it doesn't work, I've tried, I tell myself.

A Fleet Foxes song partially came back to me as I was busying with the bridge grafting – a tune playing in my mind, fragments of lyrics about working in an orchard.

I struggle to remember any other words, sing one line over again and smile. I gather up the discarded sections of apple stem.

'Back to the brambles,' I say, to no one in particular.

Though there was a wonderful delight to be had from performing this task, the truth is that there are many more hours' hard labour needed clearing bramble roots. I have sown only one area. There are seven other sections to be dug out.

RENATURING

22 March
Second section sown.

Approximately sixty square metres: a bramble-cleared area on the western side of the field.

I would like to think I'm more efficient at digging out the bramble roots, but I'm not sure I am. I have learnt to try to keep my feet secure on the ground, knees bent as I use the mattock. Yet still my body aches at the end of the day after a few hours of hacking away.

It is a fine, sunny dawn. I spend the morning in the field, clearing the last knuckles from the patch before sowing with the next bowlful of wildflower meadow and cornfield mix.

23 March
Digging bramble.

24 March
The sun rises as a fiery red ball into a clear sky a little before 6 a.m. I stand in the field and watch.

Some years ago, I was told of the Swedish word *gökotta* which means to wake early in the morning in order to go out to hear the first birds sing. Just knowing the existence of the word has enhanced my world – I picture Nordic people turning from their beds in the dark and stepping into pale, snowy landscapes, so as to listen to the angelic chimes of goldcrests tumbling from the tops of pine trees. I am sure, too, that merely knowing of the practice of *gökotta* – and having a note with the word stuck to my desk for all that time – has been an essential help to entice me up and out of the back door into the still peace of the newborn day.

For me the real reason for lifting the covers and leaving the warmth for the cold is the lure of sunrise. The songs of the robin and the blackbird, the wren and the chiffchaff are an accompaniment – a welcome and delightful one – yet it is the dawn sky that my attention is truly drawn to. There really should be a single word to describe the practice of waking early in order to see the sun rise.

We need new words to give detail, precision to our actions.
Like renaturing.

25 March
Dad's birthday.

Between Zoom calls at the kitchen table, I step out to the field and work at the next section, inching away at clearing the land of bramble roots. It acts as a physical contrast to the labour of being sat at a computer screen. Once a meeting is over, I change into my worn work clothes – muddy, torn trousers duct-taped together again, an old country shirt of Dad's and a ripped jumper. My field uniform donned, I am out the back door, over the green lane and in another world, another way of being.

It is certainly good exercise, this mattocking. Though the day is cold, within a few minutes of chopping at the ground, I am warm. I dig away for an hour then head inside again, changing back into rather smarter clothes in time for my next meeting.

28 March
A third section sown.

1 April
I wake to a thin sheet of snow. The daffodils sag under the cold weight that has been placed on them. They will be fine.

In a folder tucked away since winter, I discover a Christmas card from my friend Frances Mount which contains a packet of seeds gathered from her garden in Polstead.

'Evening primrose, *Oenathera biennis*,' says the label in her spidery, pencilled writing.

RENATURING

There's a note in the card.

'Seeds which should germinate in the spring – as promised.'

I shall wait until the snows have melted and then scatter them in the field. There is a patch of the cottage garden crowded with a collection of wildflower plants that Frances gave me back in the autumn and which now are settling into their new lands. They are dotted around my reading chair with a hope that I shall keep an eye on them and by the end of their first season will know each by name.

5 April

A blackthorn winter has sprung upon us. The hedgerows are suddenly alight with snowy-white blossom. I read William Cobbett from 1825:

> It is a remarkable fact that there is always a spell of cold and angry weather just at the time this hardy little tree is in bloom. The country people call it the *Black Thorn winter* and thus it has been called, I dare say, by all the inhabitants of this island, from generation to generation, for a thousand years.[4]

The day is certainly cold, though not angry. Blue skies frame the sudden beauty of the blackthorn which has burst out.

Yesterday, I went up to see my son Joe in Nottingham. We played in the garden, inventing games and hiding in a den beneath the trampoline, then charging out in battle, sticks in hand, shouting and screaming. He helps me to become wilder, more human. I found a magnifying glass that was intended for examining insects and the like. The sun shone. I gave the glass a clean with the end of my shirt. A memory from my own childhood returned. I collected up a few dried birch leaves, bunched them together on a slab of stone.

'Hey, Joe,' I said. 'Look.'

I held the magnifying glass, shifting it in the sun to create a tiny dot of light on the leaves. For a few seconds, we were both held perfectly rapt in anticipation.

'It's smoking,' Joe cried.

The first puffs became a distinct stream of smoke.

We gathered more material for the fire – dried moss, some small sticks, a few scraps of paper – then settled again as the sunlight focused to a bright spot that brought smoke rising, blackening the paper. We huddled silent and watched. It took only a moment.

'It's flaming,' Joe called.

'Yeah.'

We both stared at the little fire before us. Our own fire.

11 April

Mark and Selfie come over after lunch for a wander. I make tea and we step out for a look in the field clutching mugs.

'Is that the tree you've bridge-grafted?' asks Selfie.

Mark must have told him of my efforts at saving the apple tree.

'Yeah,' I say. 'Come and have a look.'

He kneels down by the tree and falls into that silent focus of the true countryman as he examines the graft. I am reminded that Selfie worked for years on fruit farms.

'You ever done this?' I ask.

'I've seen it in the orchards,' he says. 'Seen the old boys doing it. Never done it myself.'

He tells me how he's watched the operation being undertaken in the apple orchards over at Cornard by taking a sucker and grafting it across the damaged section.

'Did it ever work?'

'Yeah,' he says.

He'd seen the evidence – bridges years old successfully grafted

RENATURING

into the body of the tree. I ask if the orchards are still there, but already know the answer.

'All gone,' he says. 'Three hundred acres.'

My face grimaces. It's a common enough story around these parts.

I start to explain how I added the moss to try to keep the grafts moist.

'You got to keep them covered,' he says.

Selfie is digging his thumbnail into the exposed skin of the tree where the bark has been stripped.

'I seen a lot worse.'

He looks at the mark left by his nail. It shows a green tinge where the nail has broken the surface.

'Might be alright anyway.'

It could be that only the outer layers of bark have been lost, that the hydraulic structures of the tree are still working. Certainly, the tree seems to be doing OK. There's white blossom and the first leaves are showing.

'You'll know in about a month's time,' says Selfie.

We head over to join Mark, who's examining the CDs slung up on twine around the seeded areas in a hopeful attempt to keep birds off. They're a collection of my daughters' children's nursery rhymes and Christmas carols.

'Jingle bells?' asks Mark.

12 April

My first sighting of a swallow this year. They're back from Africa. I could cry with joy.

On 13 April 1768, Gilbert White, eighteenth-century vicar and naturalist, simply wrote '*Hirundo domestica!!!*' in his diary on seeing the swallows had returned. The three exclamation marks tell of his utter delight.

16 April

Planted the two great mullein plants that have sat in pots since last summer. Spoke to Tony over the fence who said the bloke who owned Parks Farm might well lend his digger if I wanted, for the pond.

'I'll ask him,' said Tony, 'next time I see him.'

18 April

A sunny Sunday morning. I am out early. A blackcap is singing from atop an oak branch in the field, all gusto and hope.

19 April

Time is against me. I need to get all the sections sown with wildflower seed before the end of the month at the latest. I scarify the ground with that strange three-clawed tool that I found years ago in the shed which scrapes at the surface of the soil like the foot of some prehistoric creature.

The swallow is there early morning, sat preening his wing feathers, and then gone for much of the day, which I put down to the presence of workmen digging up the road directly beneath to install hyper-fast broadband. Late afternoon, I weed the strawberry patch. The road has returned to quiet and the sun is shining. He appears again. Chittering. Busy lifting one long wing, then the other; the two prongs of his tail pointing to the ground.

My neighbours Jane and Bruce pass with their rescue black Labrador, Rocky. I call a hello, then rise from my knees and we chat over the hedge. The swallow appears on the wire above their heads.

'Just back from Africa,' I say, pointing, oddly proud.

20 April

I am running out of wildflower seed and realise that there simply isn't enough to go round for the remaining sections to the west.

RENATURING

It's something of a saving grace as I won't manage to dig out the bramble roots in time anyway. So I vow to clear and prepare one more strip down to the wild service tree, seed that and then call it a day. The remaining two outcrop patches will act as control areas – to see what arises from the soils naturally. Apart from bramble.

Having dug a new, small, triangular patch at the front of the field, I spot the remnant of a cherry tree which had been taken out of the cottage garden years earlier. Now, weathered and worn, the remains have an intriguing structure. I lift it from the ground and turn it upside down. The root stub looks like some strange bird's head, a mythical crow perhaps. I look again and wonder if in fact there isn't some Green Man face peering back at me from within the whorls and cracks of the wood. Anyhow, the tree will make a fine piece of natural art. So I dig a hole and fix the trunk at the apex of the triangle of freshly prepared earth.

There feels something deeply atavistic in this action. I think of the ancient monument of Seahenge discovered on the Norfolk coast at Holme where a great oak had been inverted and placed in the ground 4,000 years ago at the centre of a carefully prepared circle of timber. I stand back. The stump sculpture seems to fit in nicely.

I head inside to weigh out some of the precious remaining seed. Though only some ten square metres, the new triangular section will have to go light on meadow mix. I add some twenty grams of the special cornfield mix and then some of the remaining devil's-bit scabious and musk mallow. It should be a fine floral display come summer. I scatter the seed and then net the area, as it is small enough to do so, with the old net I once used to keep the blackbirds from the strawberries.

About an hour later, I spot a wren perched on a root of the upturned cherry tree.

21 April
The sun rises as a red ball this morning. I watch from the field. The point at which the sun actually appears is so much further to the north than in the wintertime. In January, I watched dawn break over Seven Acre Wood. Now, I stand beside a jay-sown oak in the field. The first light falls on me as the sun rises over the blackthorn hedge on Parks Farm Lane.

In the falling-down garden shed, a wren has made a nest in a plant pot where now a young brood of chicks rest, visible only by the glint of their black eyes. Each time I open the shed door, there is a dark flit in the far corner as a parent flees through one of the open spaces where once wooden planks formed a wall.

Later, my partner Maddie comes over. We venture to the shed, creep quietly the last couple of yards past the rhubarb patch and open the door as gently as possible, though it creaks noisily. There is chaos inside. The chicks have fledged. Half a dozen balls of fluff fly about erratically. One crashes into a plastic plant pot. Another flutters through the gap where there was once a window pane.

One simply stands on the rope of a children's swing, long since stashed in the far corner, and merely stares. I can see the strands of black fluff on the top of its head.

'Aaahhh,' says Maddie.

We cannot help but laugh at their antics. They are beyond cute. The shed may be gradually collapsing, yet it is largely dry and a pretty safe nursery for these trainee aviators.

We leave them in peace and head off to see the bluebells in Arger Fen.

23 April
Working away at clearing what I now realise will be the final section of wildflower meadow, I halt and look up. There is

something in the grasses of the field some twenty metres away. I can't make out what it is, so let the mattock fall against the wheelbarrow with a clunk and step over. It is the dried remains of a giant puff ball that I picked back in autumn from the woods down past Parks Farm. I kick it. A brown cloud of spores pours forth and drifts away on the wind. I lift the tattered husk, which feels like a dried sponge. Its fecund smell is of forest floors. I drop it back to the ground and pump it with my foot, releasing puffs of earthen air.

'May there be giant puff balls,' I say out loud and watch the smoky waves of the spores wafting over the field.

After lunch, I head off over to Two Oak Hill to collect some wild garlic for supper.

24 April
It is the weekend. I leave Molly and Eva to their own devices, enjoying a leisurely Saturday morning.

'I'm just going to the field,' I call from the back door.

Another hour spent staring at the ground. I work the soil and notice that closing-in of vision – to the details of the land before my face as my body leans over, my legs brace and my arms bring the mattock down once more.

I cannot help but think of this change in terms of the shift from the hunter-gatherer lifestyle to the lives of the first farmers of the Neolithic, who would have gazed now with anxious eyes to the earth. Their ancestors of the Mesolithic were actively looking out at the landscapes surrounding them, seeing the signs of the environment that told of the animals and the plants that lived there. The change in vision is mirrored in movement. The roaming ways of the hunter-gatherers, following the patterns of nature, the pathways of their prey, altered in the Neolithic as people started to become stationary, as they worked a parcel of land, a patch of prairie and forged their existence from the local.

My musing is over. I turn back to the land. There are still twenty, thirty square metres of the section to be cleared. Time passes. The days roll on. I need to get the seed down before the end of the month. That is the deadline I have set myself.

There is no rain. Still there is no rain.

I am starting to think like a farmer.

A few more minutes of mattocking, then I will head in and see how the girls are doing.

25 April

It is done. The final section of meadow is sown. By my calculations the area was some 88 square metres, so should have received 352 grams of wildflower meadow mix. Instead it got 125 grams – all that was left. I added 50 grams of cornfield seed as a nurse but the patch should really have had close to 200 grams. But it is done. I threw into the mixing pot all the remaining seeds I had – great mullein, devil's-bit scabious, musk mallow and cornflower – and wished them the best of luck as I scattered the last handfuls.

Now I can lift my eyes from the ground.

I watch a robin feeding a fluffy offspring. I follow the rabbit tracks and end up beside one of the young oaks that has held on to many of its dry leaves all through winter. The first fresh new leaves are starting to spring from the buds, waxy to the touch, that delicious dark lime hue. For a moment, I am held by the wonder of those oak leaves, lost from the present.

26 April

Still no rain. I step over to the field after walking Molly to the school bus stop. The ground is so hard, a pale grey on the pathways where the dry clay has turned to stone. I fill a watering can and head to the apple trees and discover that there are signs of life on the bridge grafts – at least two of the suckers have swelling buds, which means they are still alive, still passing water and nutrients through. I smile. I cannot help but reach a finger out to touch one of the buds.

Tony calls from over the fence. Somehow, I haven't noticed the dogs heading out.

'You seen the tree?'

'Yeah,' I call back.

I head over, smiling again like a proud dad.

'Buds are coming out,' he says. 'Shows it's worked.'

'Yeah.'

I tell him about Selfie coming over, how he worked for years at the orchards over at Cornard, about him digging his nail in.

'So it might have survived anyway,' I say. 'But I'm going to say it was the bridge graft.'

Tony laughs.

'I would,' he agrees.

27 April

Still no rain.

I forge a new pathway through the rising nettle patch with my steps back and forth with full bucket of water and a watering can, gradually wetting the sections sown with seed. It is a slow process. There's something oddly calming in simply going back and forth, filling the bucket and can and carrying them over to water the hard, pale ground, looking out for the green touch of life.

29 April

There are 329 square metres of the field that are cleared and sown with wildflower meadow seed. All are now watered.

3 May

Yesterday, I drove up to see Joe. He is six. We played out in the garden, the usual games that struck me once more as those of the hunter-gatherer – choosing sticks to be our weapons and then charging into the undergrowth of trees with them, hacking and stabbing away. The garden wasn't big enough for stone-throwing – another throwback to the prehistoric ways of the Mesolithic, a skill that would have been honed by the children.

Allowing ourselves to be wild again is a fine thing.

A month back, we had lit that fire with a magnifying glass. Now we set to gathering up small sticks and dried leaves once more. I had shown Joe a lighter I had bought which had an extended neck, so much the better for lighting bonfires in the field.

'Will you bring it with you?' he had pleaded.

That fascination with fire ran deep. I remember as a child being held in the same rapt wonder as Joe had been by those flames that suddenly appeared, rising from nowhere.

We shouldn't be surprised. For millions of years, our lives have been hugely enhanced, made far more comfortable and safe, by the ability to forge fire whenever and wherever we want it. The arrival of our modern world hasn't removed that most vital delight in seeing the emergence of flames where there were none.

In much the same way, those elements of hunter-gatherer living that are elicited in the games which Joe and I enacted in the garden are another reminder of the deeply held remembrances of previous ways of being that still lie within us. We stripped branches of leaves and bark to enhance our bespoken sticks, our weapons. There were other games, too.

'We need to make a potion,' announced Joe.

It was another game he'd often watched his sisters playing. He already had a jar of flour. We set to foraging around the garden –

bright-yellow dandelion heads, a handful of the tiny blue forget-me-not flowers.

'We'll leave the bluebells,' I said. 'I think they're poisonous.'

I was half-conscious of saving the few bluebells from being picked, though something told me that they were hyacinth family and so not to be eaten.

'These ones are,' said Joe, pointing to the bulbous heads of the Euphorbia. 'Nana told me.'

We busied with gathering.

I plucked a pinch of pink cherry blossom from out of the tree.

'I'll get caterpillars,' said Joe.

These were the pale brown catkins from the silver birch.

'And daisies.'

Only in the last few thousand years have we turned to the practice of growing what we need to eat on one local area. Before that we would wander the lands using the maps we had built in our heads, gathering those herbs and plants we wanted. That state of existence has not been erased from the darker corners of our heads. In the natural play of children, we see aspects of the far distant past brought back into the light. Joe and I gathered the ingredients for a magic potion and, as we did, some vestige of a time when humans were more intimately connected with the natural world, with the environments in which they lived, briefly returned – and in those moments, we worked silently away together and knew a calm, a peace in practising those ancient actions.

4 May

A huge hornet rests on the floor of my study. The creature has clearly just awoken from the long sleep of hibernation. A black antenna shifts. There is a soft, muffled sound that in time will turn into a buzz. In the field last week, a very much more awake hornet flew into my head as I was carrying a bucket of water in one hand and a full watering can in the other. The volume of the violent, drilling drone it emitted was terrifying.

5 May
There has been an odd sense of waiting over the last week. The seed is sown in the field and there has even been some rain so I no longer need to be traipsing the paths with bucket and watering can. All I can do is watch the earth and hope. There are green touches in places – signs in those first leaves and wisps of grasses that there will be a wildflower meadow of some sort by the summer.

I turn to the garden of the cottage, which has suddenly gone wild. The splash of rain meant the stubs of cow parsley shot up in a matter of hours. Over the last ten years, the area of the garden that I mow has gradually reduced to a strip of pathway to walk down across the front of the house and a patch to sit out on at the back.

Of the five centuries that this cottage has existed, it is only around the last thirty years that there has been lawn at all. The farm labourers who would have lived here had no patch of grass to maintain. The ground around their home would have been used to grow vegetables. Only a couple of weeks back, I was speaking to Diane from down the road about Mary and Bill Simmonds, who lived here for many years up until the early 1990s. Bill was a cowman to the manor. Like that long line of countrymen who'd lived here before him, he dug his home soils to supplement the household income.

'He was a right character,' Diane told me. 'Both my boys would always come and find him.'

She laughed.

'Taught 'em shooting and catching rabbits.'

That had always been a part of the country diet.

RENATURING

In the garden, there are two apple trees that would probably have been planted by Bill or Mary. They sit the perfect distance apart to hang a hammock between their boughs. Every year, they produce a prodigious crop of cookers. As I unfurl the electric cord for the mower, I realise how the entire garden would have been used to grow food. A lawn seems such a luxury. It is. The notion of the English lawn has so many suburban overtones. Somehow, it feels as though there's some link between that sterile human world and the ecocide that the lawn exemplifies. With climate change, there are now ever drier summers and ever-greater difficulties in maintaining that perfect baize. This is surely the time to shift away from seeing a 'lawnscape' as the desired suburban garden. Just yesterday, I'd seen a gardener's van parked at a neighbour's, the driver donning a chemical sprayer on his back. That is no answer. We should all leave wild places where the dandelions, forget-me-nots, cow parsley and nettles can thrive.

Later, I manage to speak to Ben Rees – the bee man. He lives half a mile or so away, just on the other side of the village, with his wife Louise, and is a professional bee keeper. He takes his hives all over the surrounding lands and harvests the honey in vast vats. Rees Bees specialises in borage honey. It is delicious – that pale yellow colour of Winnie the Pooh honey, perfectly viscous and with a sweet, silky texture. I've been meaning to ask Ben for some time about the possibility of putting one of his hives in the field once the wildflower meadows come into bloom.

'To be honest,' he says with his kind, soft voice, 'you don't need a hive.'

His bees will find the flowers, he explains. They're just a couple of fields away, after all. I can have all the benefits of having his bees nearby without the hassle of having to deal with a hive.

We start to chat about how so many people have taken to bee-keeping in recent years, which is wonderful, but how in some places this is actually starting to work to the detriment of the wild bee populations due to the competition for pollen. It is an issue Ben is keenly aware of, though as we live in such a rural world, there is less immediate concern.

'You'll be able to tell if they're honeybees or wild ones quite easily,' he continues. 'The honeybees have brown bands across their abdomen whereas the wild bees are noticeably blacker. They're the natives.'

All the honeybees I've spotted in the field will almost certainly have come from Ben's garden.

We chat on the process of creating wildflower meadows and I tell him of my hours of mattocking bramble roots. I've seen the gardens around his house when visiting to buy jars of honey. They're suitably wild. What is also noticeable is that scattered about the grounds are a number of brick blocks, each with little wooden boxes sat on top. I've asked about them before. They are part of the bee business. Each box contains a young queen. At the right moment in the spring, they will be given their own hive to start a new colony.

Ben tells me how he only cuts the grass back twice a year.

'Funny what appears,' he says. 'We had up to around a hundred bee orchids a few years back. Now they're virtually all gone. They just came and then went.'

I think of that single dried orchid in an envelope that my friend Yalda had given me back in autumn, the fine powder of seed collected in the corner. I'd scattered them in the field. Perhaps there will be bee orchids in a couple of years for Ben's honeybees to pollinate.

6 May
A day in London.

In the still of the Rare Books Reading Room in the British Library, I hold the first volume of John Gerard's *The Herball*. The book dates to 1597. A thick, black leather cover protects the pages and in the centre is an embossed gold-leafed lozenge formed

RENATURING

from an arrangement of stylised thistles, flower heads and two acorns in their cups. I open the book.

The frontispiece has the words:

The Herball or Generall Historie of Plantes
Gathered by John Gerarde
Of London Master in
Chirvrgerie
Imprinted at London by
John Norton
1597

For a while, I try to work out what that word Chirvrgerie might mean. On the opposite page there are four words written in wide-nibbed calligraphic strokes: 'Both Volumes are Perfect'.

Who wrote these, I wonder. Was it John Gerard himself on seeing this copy of the first edition? Or was it perhaps Sir William Cecil, Baron of Burghley, Lord High Treasurer of England, to whom the book was dedicated?

I turn the pages. A series of the most beautifully painted illustrations of plants sweep past my eyes. In that vast, still room, I gasp – physically gasp – at the sights before me. My fingers halt at an exquisite image. *Iris Camerarii*. Germane Flower de-luce.

7 May

As I open the door into the study this morning, a young queen is buzzing frantically at the window. At first I think it is another hornet, then realise it's a bumblebee and step over to let it out into the spring sunshine.

'Terrestris,' I say.

The yellow-banded body tells me it is a *Bombus terrestris* recently awoken from hibernation and probably seeking a suitable hole in the ground somewhere to make a new home.

Later that morning, I visit the two apple trees in the field. They are both in full blossom. The bridge grafts seem to have taken – at least two of the grafts have buds that are coming into leaf, which is certainly a good sign. In fact, the tree is looking really well, with fresh leaves and flowers. I smile, seeing the mark of Selfie's thumbnail still in the exposed section of trunk. Maybe it would have survived anyway.

There is a hum as honeybees dip from one spray of pink blossom to another. I lean in to see one bee nearest me – a distinct pattern of brown bands. It is the same with another. They have come from over there, a few fields to the west, from Ben and Louise Rees's garden. Then I see a third bee, which has a far darker shade to the abdomen.

'Wild,' I say quietly, out loud.

For a moment, I simply stand and watch that wild honeybee going about its day, flying the short trip to another bud in bloom and stepping about, extracting the nectar and tucking it away, then heading on to the next. There is something oddly exhilarating in watching this mundane activity, a sense of wonder and delight in being held captive by the motions of this wild creature.

8 May

A Saturday, though grey and unseasonal. I drop Eva off at the bus stop just as rain starts to fall. She's heading into Colchester to meet friends.

Later in the day, Molly asks if we can go and look for frogspawn.

'Sure,' I say. 'It might be a bit late for frogspawn but there could be little frogs.'

We head down the road to the pond by Parks Farm, her on a scooter and me walk/jogging to keep up, one hand on the jam jar in my pocket. I catch up on the slope by the big oak tree.

'I think we should have a pond in the field,' I say.

RENATURING

'Yeah,' she agrees. 'With frogs. And newts.'

Exactly. I can't help but think of that moment last spring when Molly fished a great crested newt out of the little pond in the garden.

'And we can swim in there, too,' I add.

It is a grand plan. The notion of a pond is lodged. A big pond. One big enough to bathe in. We hurry down to Parks Farm but there are no signs of frogspawn, tadpoles or anything amphibian.

'When we have a pond in the field, the frogs will come to us,' I say and we head home, me with images of a glistening pool of water tucked in a glade of oaks.

9 May

My friends Mark and Selfie come over after lunch. I make tea and we walk over to the field. I want to show Selfie the bridge grafting on the apple tree. He crouches down, examines the grafts quietly a moment, then rises and touches a spray of blossom.

'Doing fine, isn't it?' he says. 'You'd have been able to tell by now for sure if it wasn't getting water and nutrients up from the roots.'

'Yeah, sure,' I say.

We walk on, over to a burnt circle on the ground.

'So here's where the pond will go,' I say. 'Or, rather, wild bathing pool.'

The three of us have got previous at digging in the field. There was that day a few years back when we'd dug down some six feet on the site where eighty-year-old dowser Colin Peel had declared there was something interesting. His metal dowsing rods had certainly moved over the spot. I'd seen them. But all we found was clay – a solid continuity of clay from around a foot down.

'If I hired a mini digger for the weekend, d'you reckon that would be long enough to dig it?'

'How deep?'

'Five foot or so in the middle?'

Selfie looks at the ground before us.

'Yeah, maybe,' he says. 'You gonna have a lot of spoil to get rid of.'

I point to the end of the field, to a wooden fence that separates this wild space from the green grass of the horse-grazing paddock beyond.

'Can make a bank along there,' I say.

All three of us look to the fence, then back to the ground before us.

'It'll be a big bank.'

10 May

There's a little plant growing in the pea shingle at the back door that has just produced a single five-petal yellow flower. I use the LeafSnap app that Maddie has put me onto. The flower is identified as Blessed Herb, also known as St Benedict's Herb. I text Paul about it. His reply pings in a few moments later.

> Now if you'd said Geum urbanum I wouldn't have had to Google that!

Such is the value of the Latin binomial. But if Linnaeus only brought in the system in the eighteenth century or so, what happened before then? I text Paul back.

> Haha. And what would St Benedict have called it?

I look at that sweet, yellow flower by the kitchen door. It's an intriguing one. Benedict would have spoken Latin, I guess. But the name *Geum urbanum* would only have been given way after he was dead and buried. I wonder whether he even had anything to do with this plant. No doubt the name St Benedict's Herb was some far later adoption, probably nothing to do with the actual early Christian monk.

I can't help myself. The laptop is just there on the kitchen table. I google 'St Benedict's Herb' and learn it is best known as wood avens and grows all over Europe. There's nothing directly to do with linking Benedict to the plant but Wikipedia tells me, 'It was associated with Christianity because its leaves grew in threes and its petal in fives (reminiscent of, respectfully, the Holy Trinity and Five Wounds).' Apparently, it's also got the power to 'drive away evil spirits' and 'protect against rabid dogs and venomous snakes'. I laugh. That seems to cover most bases. No wonder it was also called Blessed Herb. Perhaps Benedict used it to keep himself safe when he was a hermit in the wilderness. I should ask my friend, the writer Sara Maitland. She would know.

12 May
Rains have continued to fall. A downpour at lunchtime yesterday filled the water butts. In the field, there is a distinct green haze over some of the cleared patches – signs of a meadow emerging. I watch from the kitchen table, flooded with essays to mark.

In the afternoon, Lynda from down the road knocks at the door. She holds out two small boxes for me.

'Found these in the garden shed,' she says. 'I must have got them a while ago. I won't be using them but I know you might be able to.'

The boxes are wildflower meadow seed mix. It's very kind of her to think of me. I thank her and tell how the bramble digging had to cease as I ran out of seed.

'Now I can get back to my mattock,' I joke, holding up the boxes.

Lynda has already given me a tray of ox-eye daisy seedlings, back in September. I repaid the favour by leaving a bag of cooking apples for her every now and then through autumn. Lynda's elderly mother likes them stewed. Neil – Lynda's husband – who I chat to whenever I see him in their front garden, brought me over half a dozen wooden pallets in his 4 x 4 when I'd asked after one left outside their house.

'They're great for drying bramble roots on,' I'd explained.

'That one's promised for Colin,' he'd said. 'But I can get you some.'

Sure enough, a few days later there was a pile of pallets resting in front of the shed.

13 May
St Mark's flies hang in the air. I watch one gradually float past – the black frame of the abdomen draped down beneath horizontal arms, remembering the tale of how Linnaeus named them after the figure of St Mark dragging the cross he was to be crucified on around the streets of Alexandria. *Bibio marco.*

14 May
The pond is begun. The spade cuts easily into the dark circle of ground where I'd burnt a vast bramble pyre back in February. There is around a foot of earth, then clay. Each slice with the spade severs another section of top soil that thumps in the wheelbarrow – grassy-haired and often still clutching a mane of ragwort. Once full enough to still allow pushing, I wheel the spoil twenty yards or so to where the first loads have begun to form a bank along the northern edge of the field. Soon there is a distinct hole.

15 May
On the drive into Colchester, I listen to a podcast from Rewilding Earth which looks to the aim of rewilding half the world. The speaker is Dennis Liu of the E.O. Wilson Biodiversity Foundation,

who is working with teachers across the US at building a community of educators framing biodiversity and conservation within the curriculum. It reassures me to know that people like this exist, that the Half-Earth Project is a thing.

16 May
A copy of *Wild Fell* by Lee Schofield arrives in the post, handed over at the front door by a new postwoman. On the cover of the book is a beautifully embossed image of a golden eagle. The blurb states:
 'In 2015, England's last and loneliest golden eagle died in an unmarked spot among the remote eastern fells of the Lake District.'
 I try to remember which year it was that I walked for miles for a sighting of what was then the last pair of golden eagles in England. I think it was 1998.

20 May
The sparrows wake me early today, urging me to rise. So I get up, dress and slip outside to the silence of dawn.

21 May
The pond in the field progresses. An hour here and there has produced a space a foot deep, six foot wide by a little more long, a gap in the greenery. The bank of soil grows. It is slow-going. I will not rush. I do not want to, nor will my knees allow me to. Each spadeful I dig out, I break apart in the wheelbarrow, peering into the earth. There are occasional flints, and balls of worms all tied up in knots that spill out. There is the odd piece of potsherd, but other than one fragment of clay pipe, very little sign of human life. One slice of the spade reveals a tiny pale lantern some nine inches or more from the surface – the paper nest of a wasp had been cut clear through by the blade, exposing the half-dozen perfectly hexagonal cells where the young had been raised.

24 May
In the clear morning sunlight, I step out and sit beside a patch of meadow. The colours of the cornfield annuals are blazing. A joy seems to burst from within me. It is hard not to weep at the beauty.

27 May
Back in March last year, I met up with Julia Boulton. Her grandmother was the renowned Beth Chatto, whose seven-acre gardens were lauded across the lands. Julia had been keen to show me something of the place but we had spent so long chatting over coffee, exploring the shelves of archives and talking with Dave the head gardener, there hadn't been time to actually see the gardens themselves.

'I'll come back in a couple of weeks,' I said.

Yet it is only now, over a year on, that I have arranged that return visit. There is fine weather forecast today, for the first time in weeks.

The car park is already nearly full.

The sense of release is palpable in the people around me. I feel it too. The sun shines. I blink in the brightness.

'Hello, again,' says Julia, who has appeared out of the sunlight.

She is smiling and wearing a suitably flowery dress.

We head to an outdoor area where the gentle hum of visitors conversing fills the air. It is eleven o'clock. Coffee time.

'There's one,' says Julia, darting for the one free table.

I follow.

'Busy,' I say.

'Yes,' agrees Julia. 'We've actually cut down much of what we were doing, catering-wise. No food prepared on-site now. All brought in from local suppliers.' She lifts a neatly designed sandwich package.

'It's so good,' she continues. 'Now we can focus back on the plants.'

People are now spending more money on plants, less on food.

'My grandma never intended this place to be one of those garden centres people go to for lunch,' Julia laughs.

There are other changes, too.

'So now you've got Chris Gibson working for you as a wildlife advisor?' I say. I've heard from Chris.

'Yes,' Julia beams. 'It's brilliant to have him.'

We both gush at what an amazing naturalist Chris is. He will act as consultant on all aspects of the gardens to ensure that biodiversity and ecological considerations are at the forefront of everything. Our chat soon shifts to other incredible people doing wonderful things for nature. Julia tells of John Little, a gardener in South Essex doing amazing things with green roofs and recycling. I mention Darren Tansley of Essex Wildlife Trust and how I've heard they have introduced beavers at the village of Finchingfield, how they are helping control the flooding there. The conversation is soon on a rather euphoric footing.

'We're going to save the world one plant at a time,' announces Julia, beaming.

29 May

Saturday. A day of sunshine. After the weekly shop, I drop Eva at her friend's house and return, eager to get back to digging the pond. Molly is happy playing computer games for a while. I get changed into field clothes. There is a knock at the front door. It is Tony from next door, holding out an old jam jar with a sticker labelled 'Nails' written across it.

'What d'you reckon this is?' he asks.

In the pot is a creature of some sort, an inch long.

'Well, it's a beetle,' I start to say.

'I've lived in the country most of my life but I've never seen one of these,' Tony states. He is clearly bemused.

It certainly is an odd-looking thing.

'Come in, Tony,' I say, suddenly realising we are still in the door frame.

He steps into the playroom. I look down into the jar. There are quite distinct antennae on the side of the beetle's head that make me think of miniature hands.

Molly comes over and peers into the pot too. We can hear the noise of scrabbling feet against the container.

'That's weird,' she says.

'I'll try to find out what it is,' I say, and take a photo then send it over to Chris Gibson.

Within a couple of minutes of Tony leaving, there is a text message reply.

> It's a Cockchafer, aka May Bug aka Billy Witch and a whole lot of other names. Grubs eat grass roots, evening-flying adults munched by Noctules.

'Cockchafer,' I announce to the room. 'Of course.'

I am sure there is a Gilbert White comment on chafers. I'll look it up later. For now, I turn to the pond.

Later that day, Mark and Selfie drop by. We sit in the garden and drink tea. It is too hot to do much more. Mark was out seeking nightingales around Polstead the night before with his farmer friend Rob Partridge. He plays an audio file of one they heard around midnight – that fluid, bubbling birdsong suddenly pouring from his phone into the garden air. When it is over he tells us how they were driving about the country lanes in Rob's Land Rover with the windows open, wind blowing.

'Then something flew into my hair,' he says, and with his hand re-enacts pulling a creature from his shaggy locks.

'You know what it was?' he asks, eyes wide.
'Cockchafer?' I say.
'Yeah!' Mark says with mock annoyance. 'It was.'
I tell them of Tony coming by earlier.

Early the next morning, I write a note to Tony, telling him what Chris Gibson told me, and pop it in his letter box. It feels rather fun passing a message on in this way, telling tales of the creatures that live in these lands alongside us.

31 May
A post by Alexis Nikole Nelson (aka @blackforager on TikTok and Instagram) explains how to drink nectar from a honeysuckle flower. These are the kinds of things that should be on social media. I watch the reel over and let it roll again.

'Happy snacking! Don't die!' she calls joyfully.[5]

1 June
Scrolling my photos on my phone, I see a picture of that cockchafer Tony brought over the other day. I turn to Gilbert White. On 9 May 1774, he wrote that 'Chafers have not been plenty since the year 1770' but that now there were 'Chafers in vast numbers'. He also wrote in his diary that day how, 'The caprimulgus is the last bird of passage but one: the stoparola is the last.' By caprimulgus he was referring to that weird and wonderful bird, the nightjar – the term meaning 'goatsucker', from the old belief that the birds surreptitiously took milk from the teats of goats and cows at night.

The reference reminds me of a fabulous venture to the Brecklands of Suffolk a few years before with my good friend Ben Castell. On a moonlit midsummer night, we drove out to those empty lands and waited and waited and finally got to watch those strange shadowy creatures flying about. At midnight, stood on a sandy heath, we saw the dark patches of nightjars flitting across our vision and heard them clapping their wings together to make the eeriest of sounds. It was a noise once heard, never forgotten.

2 June

Apparently, 'stoparola' is an archaic name for a flycatcher.

3 June

In the post is a credit-card sized token made of wood that is a Friend of Beth Chatto's membership card. An accompanying leaflet explains: 'In 1960, she began transforming an overgrown wasteland, comprising poor soil and boggy hollows, into a series of informal and tranquil gardens.'

4 June

In the field, one of the two great mullein plants is showing signs of shooting skywards. Hopefully, a passing mullein moth will find them.

Great Mullein

7 June

Frances Mount comes over for coffee, kindly transported over from Polstead by Dave and Mandy Charleston. Frances arrives clutching an offering – a yogurt pot of soil with some leaves sticking out, wrapped in a plastic bag that once held bread.

'It's a valerian,' she says.

Mandy has brought cake. I return to Dave the copy of Gary Snyder's *Good, Wild, Sacred* that I have been cherishing for some months.

We potter into the field together and chat about wildflowers.

8 June

As I drive up the lane towards Hill Farm I am trying to work out exactly how many years it has been since I last visited, but am distracted by the swaying flocks of ox-eye daisy on the sloping verges. Six, I reckon. I park and step out of the car to find Ashley Cooper already standing there. He is taller than I remember. I feel myself pull up to my full height.

'Hello,' he says very warmly, removing his floppy, wide-brimmed hat.

Only an hour earlier I'd been re-reading Jack Lindsay's *The Discovery of Britain* to jog my memory of Roman Gestingthorpe. It was Ashley's father, Harold Cooper, who had discovered the Roman site when deep ploughing back in 1948, bringing up a mass of red tiles.

I remind Ashley of his appearance in Lindsay's book, as a small boy digging away alongside his father, imagining a stone is a coin.

He smiles with the remembrance.

'Yes, indeed,' he says, and then quotes. '"He plucks poppies and frets."'[6]

I laugh. Here is that child over sixty years later. Dressed in countryman checked shirt and brown jacket, now he is in charge of farming these lands – and overseeing the ongoing archaeological investigations into the Roman site. We head off in Ashley's bright red Range Rover. By my feet are two pieces of tile. On the dashboard is a section of glass that I cannot resist examining. But I have visited not for the past, rather for the future. My friend Ellie Mead told me a week or so before that Ashley has been busy putting a series of ponds on his land. I want to know more about them.

We bump over the dry ground, down to the Sudbury Road, then take a track beside a footpath fingerpost by a blackthorn hedge and come to a halt in a grassy field interspersed with well-spaced trees. Each has a protective guard – noticeably not a plastic one – and a neat blue collar to stop the bark rubbing.

Ashley gets out. I follow. We walk over the rough path and Ashley explains how much he enjoys the physical work of caring for the willows, cutting their side branches away, ensuring they grow tall and straight.

'Modern farmers don't actually do much hard graft anymore,' he tells me. 'There's lots of sitting at computers and in tractor cabs.'

'With changing agriculture after the Second World War, there was greater emphasis on arable land, with less profit from cattle. Lots of farmers had bits of meadow they couldn't sensibly use – growing cricket-bat willows became a worthwhile option.'

Each cricket-bat willow takes about twenty years to grow.

'I planted my first ones when I was seventeen,' he says with a laugh. 'But due to the spread of watermark [a bacterial disease of willow trees], from about fifteen years ago, I was beginning to have an increasing area with nothing on.'

We arrive beside one of the ponds that Ashley has been constructing recently. He starts to tell me how these low-lying meadows across Essex and Suffolk would have been grazed by cattle for centuries.

Ashley is a thoughtful, conscientious farmer. Like so many

others, he is driven not by profit but by a passion for nature. He tells me of FWAG – the Farming and Wildlife Advisory Group.

'It's a farmer-subscribed organisation – no funding of any sort other than the farmers who pay into it. They have advisors who are passionate about the environment and can help you to make your own farm more beautiful and environmentally friendly.

'We applied to put in four ponds and a scrape – a mini-pond with the objective that it should dry out during the summer as the fluctuating water table and eventually degeneration to mud allows certain insects to lay their eggs in the mud before it fills with water again.'

These ponds had actually been dug, thanks to funding from a scheme under Farming Wildlife, sponsored by developers who have to pay money to Natural England if they want to develop land where there might be great crested newts nearby. Natural England then recycles it via FWAG to get other ponds dug.

'So these ponds are a combination of both government and developer funding . . .' he pauses and smiles. 'And a huge amount of input from myself.'

We both laugh.

'I'm sure,' I say.

I think of the feeble hole half begun in my field. The pond beside us is a lake in comparison.

'There are actually quite a lot of great crested newts around my way,' I say with something of a conspiratorial whisper. We stroll off through the high grasses towards the next pond. Ashley is a natural conversationalist. His intelligence and humanity seep through. He asks about my interest in the ponds.

'I'm interested in the way individuals can be involved in bringing back nature,' I say, and tell him a little of my efforts in my two-acre field.

'Not everyone has 4,000 acres around Knepp Castle,' Ashley jokes.

We chat together about that gulf between the incredible rewilding project which is taking place on the Knepp Estate and smaller, individual efforts to renature private gardens, even window boxes.

Ashley takes a breath then continues seriously.

'I think we have interfered with nature so much, so now we can't stop interfering. Rewilding has to work on an enormous scale, with apex predators and all the rest of it.'

We arrive at another pond.

He explains how the pond is about a metre and a half deep in the middle. It is at natural water-table level.

'The Belchamp Brook is wending its way over here,' he says.

'So is this what's called a "puddled" pond, then?'

It is a term I've heard before – something to do with the way in which a natural pond can be simply formed by making a hole and then using clay to keep the water from draining away.

'No, not really,' Ashley explains. 'Because we're at the natural water-table level, the digger kind of ran about a bit in the bottom and flattened it down, but it's not a genuine puddled pond in the sense someone has put six inches of clay in and "puddled" it. If the Belchamp Brook goes down, so does this.'

I start to talk about my initial efforts to dig the pond in the field.

'There's about a foot of soil and then straight into clay, so I'll have to puddle it.'

'You'll need to,' Ashley says. 'Because you're collecting water, and storing it there.'

Such is the variation of ponds.

We chat away. There is a bucolic wonder to the landscape around us – the sense of the gentle pace of the place is so clear. The ghosts of the generations of cows that once wandered these meadows can be heard chewing the cud. Songbirds sing. Electric-blue damselflies and hawker dragonflies dart about us from reed to grass head.

Every year some of the ponds will have a sample of water

taken, and from the environmental DNA (eDNA) within those samples, it is possible to tell if great crested newts are present. The signs are good.

'From the very first day when the digger driver was here, Lucy, the FWAG advisor, started talking to him, and you could see he sort of thought, "Here we go again. Another hi-vis jacket and yellow hat. Great crested newts. Heard it all before."'

Ashley jumps into life. 'And suddenly Lucy shouts "Look! There's a great created newt!" She picked it up and that chap just changed in a moment.'

'Wow.'

'He was nearly retired. He'd never seen one before,' Ashley continues gleefully. 'Lucy put it in his hands. He was no longer sixty-four but seven and a half years old. From that moment, it was no longer a job. It was suddenly something he was really interested in. A magical moment.'

'That's absolutely it,' I add.

'I said to Lucy "You really want to have one or two in your trouser pockets . . ."'

I laugh.

'They're one of those special species, aren't they? They'll stop building sites going up.'

I start to tell Ashley the tale of Molly pulling out a great crested newt from the little pond we'd dug in the garden a year before.

'It was literally her first scoop with the net.'

The remembrance of that moment: Molly's urgent call from outside; the sunshine as I ran out and the sight of that magnificent creature – her size, the brilliant blaze of her underbelly, her dragon-like crest, the padding feet, the cool touch of her skin on my hands.

'Your daughters will never forget it,' Ashley says.

'It was quite amazing,' I say. 'I think I was more excited than them.'

While there may be sightings of great crested newts in our local area, nationally they are rarely to be found. There are nine recently dug ponds here and each will hopefully offer a new habitat to these incredible beings, such that they may flourish even further in these parts.

We walk on. Of course, the ponds will provide homes to many more creatures than just great crested newts.

We arrive at the largest of the nine ponds, which is some fifty yards long by about half that across and surrounded by reeds and tall grasses blowing in the wind.

'This is lovely,' I declare.

The pond had been dug over a decade ago.

'This is where we found rather an interesting archaeological find,' says Ashley. 'We found an aurochs bone.'

'Did you really?' I exclaim rather loudly.

Aurochs were the huge, lumbering wild oxen which the first farmers domesticated and from which modern-day cows are descended. They've always held something of a fascination for me.

'Yes, just one,' Ashley says. 'At the very bottom of the dig.'

He tells me that there was also a deposit of fire-crazed flint found close to the aurochs bone – a classic tell-tale sign of prehistoric activity, with the flint heated in the fire to then be used to boil water.

A hawker dragonfly hovers past. Swallows and martins slip through the sky above. It is easy to dip into imagining the scene of Bronze Age or Stone Age people stepping on these same soils as they come to the fresh water of the brook just a few metres away.

'In some places, there is a wonderful waft of mint,' Ashley says.

'Water mint?'

'Yes.'

We halt and sniff the air for that so-distinct scent. I smile and think of others coming here long before to gather water and mint, and in the still of that space realise that there is something truly Edenic here.

I step back into the present and we chat about the singular impact that ponds can have on a landscape, whether it be a tiny back garden or on a grander scale such as this.

Ashley agrees. Alongside these larger ponds partially funded by FWAG, he's been busy elsewhere.

'They've encouraged me. But I've also dug them at every point on the farm where there's some confluence of water. Some of them – just like your pond at home – are no more than two or three metres across. But the benefit . . .'

'It is incredible,' I nod.

'Damselfly,' whispers Ashley suddenly, and we are both caught by the flittering flight of a slight creature that alights on a tall grass head. I lean in, peer closer. There are dark patches on each of the wings. The entire body is a stunning metallic blue-green.

'Beautiful,' I say quietly and break the silence.

Later, I will discover that the damselfly was a male banded demoiselle (*Calopteryx splendens*), which would have been performing that aerial dance to court the green-bodied females of the species rather than for me and Ashley. In that moment, held in the spell of the present, there is an incredible sense of being immersed in the vibrant life of the landscape. All around us, I notice there is the movement of insects, the hum of hundreds of beings going about their day, and to be human feels rather like being a cumbersome, clumsy giant, stomping noisily about the delicate landscapes of others – a little like those ancient aurochs perhaps.

At times, we humans can be crass, bumbling oafs, but when we think and deliberately act to carve out whole worlds from the lands for other living beings – like these ponds – we become something more: guardians, overseers, able to see beyond our own selfish, venal desires to a greater good, a caring for the other living creatures who we share this earth with and, so too, a higher worth to our own existence. So long as we don't over manage. Help restore to more 'natural' landscapes; then stand back and leave well alone.

I wander with Ashley back through these idyllic meadowlands to his red Range Rover and we head slowly away, bumping off up the gentle incline along the field edge. I think again of those Neolithic farmers clearing these lands of trees 6,000 years before. Looking up to the line of the hilltop through the splintered sunlight it is not hard to imagine their figures silhouetted, still just about able to be made out against the skyline on this perfect English summer's day. If we could just halt and learn a little from their ways, I think, we would all do well.

They are there beside us, too, as Ashley and I walk the lands by the Belchamp Brook, in those deliciously cool meadows that would have provided such rich nourishment for the aurochs that those first farming folk nurtured. The wet, marshy glades are places where not only damselflies prosper. Allowing streams and rivers to be more carefully cared for, creating ponds and scrapes where water can gather, are vital ways to renature our world. We now know something of how much those prehistoric people venerated the waterways: how in the Bronze Age they placed ritual deposits of swords, shields and precious metals at the confluences of rivers, at crossing points; how those same people built round burial circles close to the river's edge, so as to place their recently departed loved ones by the riverbank, at the height of the floodline, where those same rivers could be forded, could be bridged. It is time to learn to worship these watery worlds once more.

On Ponds

Adding water to any patch of ground is often said to be the single most effective way of enhancing biodiversity. The size of that pool of water will obviously affect the variety and scope of the plants, insects and birds that are attracted, but even the smallest washing-up-bowl-sized pond will be welcomed by some creatures. In the last century, 70 per cent of rural ponds have been lost from the countryside due the pressures of intensive farming, development and general habitat loss. Though there have been efforts by many fine farmers to restore and recreate large ponds on their lands, the importance of garden ponds has never been greater.

When I was twelve years old, our year group were given a week off school to do a project of our choosing. I did mine on pond life. As such, I spent a heavenly few summer days sat beside the pond that my dad had recently dug in our suburban back garden in London, logging the various sightings of dragonflies, newts and hoverflies. It was bliss. Being still beside still water is a simple and wonderfully effective elixir for many ills. Rather than being imprisoned in school, I could pass the days sat gazing into the pond and discovering some of the wonders that called it a home.

Building a pond is also a fabulously fun activity. Some years ago, inspired by my dad's efforts, I built a pond in the back garden with Eva and Molly, then aged six and three. Using a variety of implements – spoons, spades, sieves were all involved – we worked away, collecting a box of 'treasure' consisting of bits of pottery, clay pipe and chalk from the gradually widening hole in the ground. It was slow but steady progress as all three of us happily dug down into the earth. After a few happy days of effort and dirt, we had a space in the garden some six feet across by three feet wide – a pond, or rather, a waterless hole.

For the liner, I simply bought a roll of flexible butyl and laid that tight against the soil. Back in our suburban London garden, my dad had used fibreglass, layering sheets of the stuff on the sides of the pond and securing them with a glue that I can still smell today. Both he and I laid whatever pieces of stone we could find to form an edge – I had turned to the local fields and carried back armfuls of flints that created a rockery of sorts.

Then, to turn a hole to a pond merely add rainwater and wait.

One of the most remarkable things about building a pond is how quickly the wildlife arrives. Within a couple of days there will be pond skaters dancing over the surface of the water. Diving beetles will soon follow. Amphibians may well appear in the next few months, depending somewhat on the season. There will be water scorpions and nymphs beneath the surface and hoverflies, damselflies above. Frogs and toads need water for their frogspawn and newts of all varieties need water to breed in. You may even get lucky enough to find a great crested newt in your pond, as Molly did on a pond-dipping exercise during lockdown. It remains one of my most delightful moments of natural wondering – a sunny summer's day and Molly's face as she showed me and Eva what she had found in the pond we had dug ourselves.

Of course, if you have more land, then dig a bigger pond. The variety of wildlife the pond supports and attracts will certainly depend on the size of the pond. Even my dad's suburban pond was big enough to be visited by grey herons, who would eat every fish in sight. For a time my poor father battled to save the population of goldfish and carp, but in vain. Perhaps it wasn't a bad thing for my pond life project – fish in a garden pond not only act as top predators, eating all the insects and invertebrates they can, but they add to the nutrient levels, making algae and blanketweed more of a problem. Early on after construction, Dad and I released some sticklebacks we had caught in the Grand Union Canal into our pond. We regretted it. Years later, they were happily still there, evading all our best efforts to catch them.

An essential when creating a pond is to see the space from the eyes of those who will live there. Steep sides are no good to a frog. Gentle, sloping entrances to the water allow access for all those newts and other amphibians who like to step in and out. Adding a collection of stones and pebbles to one side of the pond will allow birds to come and drink or bathe – perhaps even a hedgehog. A smaller mosaic of stones allows honeybees, wasps and hornets to settle and drink safely without falling into the water.

Plants – both on the edges and within the pond – provide not only food and shelter, but essential oxygenation of the water. Some also offer pond inhabitants shade from the summer heat. Water lilies, with their umbrella-like leaves, are perfect for this, and there are miniature water lily varieties for a mini-pond. Marsh marigold, water forget-me-not and lesser spearwort are wonderful for providing some colour to the pond, while submerged plants like water crow-foot, water violet and hornwort will keep the water healthy. Duckweed is a plant to be avoided. Wash any incoming plants to try to keep it from creeping in on them. If algae starts to build up, suffocating the pond, a wad of barley straw held down with a stone does wonders at clearing the water. Plants are best set in containers such as aquatic baskets, with the soil held in hessian bags to prevent it washing out. This will keep the plants from spreading and overtaking the pond – something well worth avoiding if you're upcycling a washing up bowl!

In essence, there is nothing you can do more for the nature on your doorstep than creating a pond of some kind. The knock-on effects for biodiversity are huge. Be it made of an old sink or wheelbarrow basin placed in a suburban garden, a large pot on a window sill or a large rural pond in a field, that body of water will probably be the best thing you ever do for wildlife.

Whatever the size and scale of the pond, it will evolve and change through the seasons. Different creatures will come and go. Life will flourish in spring and summer and die down again over winter. And you will be able to witness all that wonderful sound and fury close up.[7]

9 June
I have discovered *Rewild or Die* by Urban Scout (aka Peter Michael Bauer), with the subtitle: *Revolution and Renaissance at the End of Civilisation*.

Rather than merely seeing rewilding as seeking large-scale ecosystem restoration, Bauer explores the ways in which we can all, as modern humans, 'rewild' if we recognise the value of hunter-gatherer lifestyles, of indigenous ways of being, if we shift our mindsets from those of the domesticated, unsustainable practices of western civilisation. This is my kind of book.

11 June
Purple-flowered barrels of selfheal have suddenly arisen in the field. For a brief moment, when I saw the first group under the leaning wild service tree, I thought they were a clump of bee orchids. My heart flew. They were not. But they are beautiful anyway and actually do have a certain bumblebee shape to their heads.

13 June
Awake at dawn.

In the field, there is something going on with the rooks. They have been doing this at dawn for a while now: congregating, filling the air with their flinty calls. The other morning I stood here and listened to them, exactly as I am now. Small groups, like family units, cluster in the tops of various oaks. Then they seem to all come together. I can't see the collection of corvids, the gathering. They are on the other side of the largest field oak, in the reservoir next door. I can hear them though. A rising collection of voices. Then there is a sudden explosion of sound – a concatenation of bird calls and a sudden, collective action: a black ball of smoke that rises into the sky and then breaks.

I count some thirty-five individual birds as they fly off in small packs. As I stand and watch, it is hard not to see the event as a morning meeting of some kind. Those birds seem to be gathering in a way some distant human ancestors might have done, tribes of hunter-gatherers that would meet at the rise of the sun, to share tales, news, plans, and then head off on their separate ways. And as I stand in the field, watching the last wisps of smoke flying away, I so wish that the ancient arts of augury – of divination, seeing truths through the actions and movements of birds – as practised by the high priests of ancient Rome, Greece and other civilisations of the past, were not lost to me, indeed were not lost to the entire modern world – even if they have been replaced by such practices as science, meteorology and climatology. Perhaps that clutching for the threads of augury is at the heart of why so many humans still practise so ardently and intently the pastime of bird-watching.

The jackdaws and the rooks have quietened, leaving the blue space of dawn to the songbirds. Now the air rings out with their high cries rather than the stony cracks and blows of the crows. Only, I do not know what any of it means. Hopefully, the auspices are favourable.

15 June
While speaking to Maddie on the phone first thing while stood in the field, I am distracted first by a marmalade hoverfly sat on a yellow head of ragwort and then by the red wing of a cinnabar moth. She is forgiving.

18 June
I first heard of Ray Davis through my good friend Matt Mackman, an artist by trade, who told me he was working on a project using birds' feathers. Matt told me he had a source for plentiful supplies from someone called Ray, who owned a few acres overlooking the Stour Valley. It sounded like a similar set-up to my field in that he was actively seeking to help bring back as much biodiversity to that land as he could – except Ray was using more direct interventions.

'He shoots magpies,' Matt told me one night over a beer.

Ray sounded like someone I needed to meet. He was clearly renaturing his patch of land in his own distinct way. Then Matt's mum, Ginny, got in touch. Along with the promise of a consignment of six Kentish cobnut seedlings for the field, she sent me Ray's phone number.

'He likes to keep himself busy, mostly by distributing venison,' she texted.

I meet Ray in the rain of a damp June day, after following his directions and turning up a side track off the Bures Road. He is sat in the shelter of his Land Rover. It is hardly a day for hearing the songbirds but at least we can walk around the five acres of land called Kinton that Ray looks after as a nature reserve, actively working to ensure the best possible environment for both the wildflowers and most of the wild creatures that live there.

RENATURING

'I control the vermin,' he states. 'You shouldn't hear a magpie. In fact, I had the trap up here last week, heard one, set the trap and caught the pair.'

Ray speaks with a touch of the Suffolk countryside in his voice.

'And grey squirrels,' he continues. 'I trim those up.'

He is quite smartly dressed, in countryman shirt, covered from the rain in a dark green jacket and cap. We've already halted our tour of the land as I want to learn more about how Ray goes about things here.

'I'm not fond of jays, actually,' he says. 'Some say, "I like the jays, they plant acorns", but they're a magpie really.'

'They're the same family, aren't they?'

'Yeah.'

'They take a lot of young chicks.'

I remember the sight of two jays terrorising a blue tit nest a couple of years back. Not that I'd ever shoot anything.

Ray plays an active role in overseeing this patch of ground. He is known around these parts as a good old country boy, one of a dwindling few whom the police will call to come out at all times of day or night to dispatch an injured deer that has been run over on the roads. While not officially a gamekeeper as such, he is respected by that tough collection of folk who supervise the land. He has spent all his life in these beautiful borderlands between Suffolk and Essex, knows all the farmers and their ways.

'I have buzzards up here, and kites,' he says. 'I feed them with the deer remains. The gralloch.'

'The what?' It is a new word for me.

'The guts of the deer.'

'Is that what you call it. Gralloch?'

'Yeah,' says Ray. 'Sometimes I bring a deer up here and hook it on the side of the Land Rover and rip the thing open. The guts, the lungs . . .'

'So they can get into it easier,' I suggest.

'Yeah, and the whole thing goes. In no time, the buzzards, the kites, the badgers, the foxes, they rip into it. You can see crows come down, sometimes a herring gull. Everything gets eaten. Nothing gets wasted up here. Any butchering I do – I knock a sheep over for the farmers, and various other bits and pieces – they all get left out here and they all disappear.'

We head off, stepping a path through the young wildflowers of the meadow areas as the rain falls. Ray's way is a very hands-on approach to managing the land. He's a countryman and one who shoots certain species for the benefit of others. It's not my way but it's interesting to hear him talk about how he acts as guardian overseer of this patch.

'No one I know manages their land as actively as you do,' I say.

Ray targets what he calls 'the vermin'.

'With the grey squirrels, don't you just get another one come in?'

'That's right,' agrees Ray. 'It's a vacuum.'

He tells how he has a trap where the squirrel goes in after a nut.

'They touch a plate and it gets them right behind the napper. Kills them instantly. I get two at a time.'

Ray has kept a record over the last year.

'From January to January, I caught a hundred and thirty squirrels. Some I shot with a .410 out the window on that feeder over there.'

He points towards the shepherd's hut behind us.

'Most I caught in the traps.'

In the time since, the numbers have been substantially down, though Ray is still catching them.

Ray's efforts to control the squirrels and magpies really intrigues me. Yet he is nothing like the person I had imagined. In truth, I had been rather expecting him to be a little more reserved, more aggressive perhaps, more macho. Instead, he is a very sweet, kind man. The fact that he uses guns and traps to kill those creatures which would otherwise overrun these five acres doesn't make him a monster. It makes him an active player in the dynamics of the food chain. Some may argue he is playing God, but maybe that is what is needed. When seeking to create ecological restoration on areas of land that aren't thousands of acres in size, how is it best to oversee the balance of creatures? Maybe Ray is acting as the missing apex predator over these lands, the wolf, the lynx that keeps the ecosystem healthy. It's not an approach I'd ever take but it is easy to see the effect that Ray is having by controlling the populations of magpies and grey squirrels. Even in the pouring rain, the air is full of birdsong.

We walk on and turn our attention elsewhere. Ray isn't merely concerned with shooting and trapping squirrels and magpies for the care of the songbirds, he's also keen to create something of a wildflower haven. Already today, I've been swooning over the swathes of yellow rattle. He gets various orchids in the meadow patches and other delights, too. A local wildflower expert has identified one wonderful-sounding delight.

'Ploughman's spikenard,' Rays says with deliberate precision.

I repeat the name. It's certainly new to me.

'Got a leaf like a foxglove,' he adds.

A visiting group of naturalists had been over a while back to do a survey.

'They said, "They're foxgloves",' he says with a smile. 'Well, I thought, I'm not going to argue with you, but they're not!'

We laugh together.

Some plants such as the ploughman's spikenard arrive of their own accord. Others, Ray brings in as seed.

'When I was down in Cornwall, I got some of that red campion. Chucked it about. Had a lovely show of campion down in that bottom bit,' he says. 'Thought to myself, "Isn't that lovely." Came up the other day and it had all been chewed off.'

He pauses.

'That's the muntjac.'

We halt beside a patch of bramble. Rain falls from above. Ray ducks down into the scrub, pulling back the greenery to reveal a mass of bones.

'I had to put down several deer last week and the whole lot are gone,' he says. 'Even the skins are gone. They're feeding the cubs, I guess.'

He's talking of the foxes and the badgers that also live on these lands. They're very well provided for by Ray. They must be thriving. He's already told me of watching the young fox cubs playing from his shepherd's hut. For those creatures, there can be few finer places to be brought up.

Ray is bent over, studying the matrix of the wildflower meadow that is still in the early stages of spring growth.

'Look at these little oaks,' he says.

'The sad thing is that they're not going to make it,' I say. It's exactly the same in my field.

'No, they're no good there,' he says.

'I've dug a few up and moved them,' I confess.

'You've got to get them early,' says Ray with an obvious sympathy. 'With that tap root.'

He says how he's had primroses and cowslips all over. I lean closer in.

'You've got such a mosaic here,' I say. 'That's all yarrow.'

The finely detailed filigree of the leaves is obvious.

'That's all wild strawberry, isn't it?' I ask.

'Yeah. They cover everywhere,' says Ray. 'And they're nice to eat. About the size of the end of your finger.'

He holds the tip of his left little finger with the index finger and thumb of his right.

'With a drop of moisture they'll plump up.'

The rain is steady now.

'This'll help them,' I say.

As we step on, I want to know more about how Ray deals with the muntjac that munch the wildflowers here.

'You just have to shoot them whenever you can?' I ask.

'Yes,' agrees Ray.

'But they'll keep coming in?'

'Oh yeah,' he says. 'They'll keep coming in. Muntjac are everywhere. They breed all the year round.'

He tells how in the wintertime he had a couple of friends come up with a dog.

'I don't like to bang about here,' he says. 'It's a lovely quiet place. But we tapped through and I stood back with the gun. Had one or two muntjac as they came through. Interesting to see what else came out. Foxes came through, rabbits, there were several woodcock . . .'

'Woodcock?' I say, interrupting. 'That's lovely.'

'I mean, I don't shoot any game up here,' Ray explains. 'I've got a pheasant up here. He won't come out now. But when I'm alone, he walks around with me. Cocky, I call him.'

He smiles and makes a noise imitating the pheasant's chatter.

'Worp, worp, worp worp.'

I laugh.

'He has a handful of sunflower hearts. There's two hens – Little Henny and Great Henny. They come round with me,' Ray says. 'But they're nesting at the moment.'

It's easy to imagine him pottering around with this troupe of pheasants following.

'There's a lot of game up here,' he explains. 'When you're in the hut, you haven't got to wait long before a lot of things come out.'

'This place must be a refuge for them,' I say.

'Yes, they're safe.'

'Exactly. They are safe. No one's going to shoot them here.'

There's an element of Eden to these five acres for these pheasants that have so fortunately found themselves living here. They wouldn't have to venture far from this patch before they would be seriously at risk of being shot. Much of the surrounding land is owned by farmers who make a living from rearing and feeding pheasants so that they can be gunned down by men willing to pay hundreds of pounds each for the privilege.

Ray is explaining how he gathers up piles of garden rubbish from people down the road in the nearby village of Little Cornard – grass cuttings and the like – that he then brings up in his Land Rover to create varied habitats for creatures such as voles and mice.

'You can see I didn't cut this patch,' says Ray. 'There was a lot of knapweed that I left for the goldfinches. A lot of cock's-foot that the voles like – they can get under that where they're safe. But you can see, all the trees come up here – just in a year.'

He points to the young hawthorns that are rising from the ground.

'If you left this, two or three years, it would be like that,' he says, looking to the scrubby patch beside us.

In the autumn, he'll cut this patch back, leave the clumps of grass.

'It's a nice bit of habitat,' he says.

'Do you get much ragwort?' I ask.

'Oh yes. I love ragwort. I encourage it.'

'Good,' I say.

'Get those cinnabar moths and caterpillars . . . I love it.'

It's lovely to hear.

'Several farmers have said to me, "You want to get rid of that." I say, "Bollocks!" It's a native plant and hosts about a hundred and fifty insects. So I do like ragwort.'

'Good on you, Ray,' I say. 'Keep telling them.'

We talk about how horses avoid it in the field. It's only if ragwort is cut up in the hay that there's a danger of them eating it. Once more, though, for me, it's so reassuring to hear Ray saying such things. He's a proper local countryman but he's wise enough to see through these myths of the countryside that are badly affecting the ways in which the natural landscape is managed. Ragwort is a valuable wildflower for so many creatures. That's the truth. The fact that Ray is telling local farmers this is by far the most impactful way of shifting their beliefs and practices – hearing it from the horse's mouth, you could say.

Common Ragwort

We walk on, stepping over clumps of knapweed on a bed of yellow rattle. The rain continues to fall. It might be June but it's certainly not a great day to listen to the songbirds. Most are tucked down undercover. We shelter, too, beside a vast pile of stacked wood. It must be ten feet high and thirty feet across. There will be a host of animals that call it home, thousands of insects that live within the structure.

'It's what I call a tidy mess,' says Ray.

He points out a log lying on the ground.

'This is where the foxes and badgers come through.'

You can see the scratch-marks in the wood from their claws. This is a trail between their homes and the open land where Ray leaves out the deer carcasses.

'The vixen will drag a whole ribcage over. They'll chew every bit of meat off it.'

The footpath leads us to a lower section of Ray's patch, pitted and varied as it was once part of a sand and gravel extraction industry that ran right along the river valley. There are all kinds of wildflowers growing – most brought in as seed by Ray. Buttercups. Comfrey. Campion. Foxgloves. Violets.

'These have suddenly come up,' he says. 'Ox-eye.'

He's also planted thousands of snowdrops. It's part of a long-term project to cover the place with flowers.

'A lady – Sylvia Roberts, who farms on the Colne Road out of Bures – she said, "Got some bluebell seed for you, Ray." I thought, "Bluebell seeds? They'll take years to grow." They're like little black onion seeds. So anyway, I threw them down – maybe a bit thicker than I should have done. The next year, they came up like little blades of grass. They take about three or four years to flower.'

Ray says how he also goes around various bluebell woods in June with a bucket.

'When they're just right. I tap the seed heads. It's a back-breaking job. You get about that much in the bottom of the bucket,' he says, indicating with a finger.

He has been doing that for three years now, spreading the seed all over the banks in this more sheltered, woody area. You can just make out the young bluebell plants as blades of green, single leaves emerging from the earth.

'See that,' he says, looking closer at the bank. 'Those are last year's.'

'You're creating a whole bluebell wood. You're just helping it to come along a bit quicker,' I say.

Local farmers have offered clumps of mature bluebells that were due to be sprayed off for pheasant penning and lost.

'So you save them from other woods and bring them here?'

'Yes,' agrees Ray.

'You're creating a refuge. Not just for animals, but for insects, plants . . .'

'Everything,' says Ray.

That sense of building a place of safety is so evident. It's the single philosophy of protection that fills Ray's energy for looking after and managing these five acres. He is a quite remarkable man. For fifty years he worked as a driving instructor, teaching most of the youth around these parts to drive, but he's always been a good old country boy.

'I always kept a gun in the car when I was teaching,' he says. 'Often used to come across a deer that had been bowled over by a car, or have a shot out the window at a squirrel or a magpie. All of the farmers knew that if they saw a white car, that was me. It was not someone poaching.'

His eyes sparkle as he remembers.

'There was a line of oak trees up at Acton on a hill, where I'd get them to do clutch control. I always had a sun roof. I'd say, "Just creep along nice and steady." I'd get the roof back. There'd be a magpie up the tree.'

I raise my eyebrows.

Ray imitates the chuckle of the magpie.

'Ick ick ick ick . . .'

Then leans back with an imaginary gun.

'Bang!'

He laughs.

'I had one that fell on the bonnet. This boy said, "Do you normally shoot magpies in a lesson?" I said, "If I get a chance, I do."'

Ray isn't finished.

'I got two in one lesson. We were on the aerodrome and this boy had never driven before. He was only going along in second gear, nice and steady. There was a great muck heap, always magpies on that. I got the .410 out the window on the passenger side – we were going nice and slow – and I went "Bang!"'

I can't help laughing.

'Got that first one. Then when we came back about ten minutes later, there was another one there, so I got that as well.'

21 June

Finally I manage to track down someone my friend Paul suggested I should talk to, who has moved in next door to him in Alphamstone. Anna Beames is her name. She works for the Suffolk Farming and Wildlife Advisory Group and is a soil expert.

Late in the afternoon, she texts to say she is around if I am free to chat. I am. It is a gorgeous day. I ring from the field and a bubbly voice answers.

'Is that Anna?' I ask.

It is her. She apologises for having been too busy to meet that week and proceeds to speak with such insight and wisdom on all aspects of landscape management, rewilding and farming. I sit on a fallen elm bough and listen. Her enthusiasm and passion are evident. She is keen to tell of the many farmers she knows who are actively experimenting with ways of improving the soil and increasing biodiversity on their land purely from a desire to better the world.

Anna speaks of the 'deep integrity' of so many of the farmers she works with. I think immediately of Ashley Cooper over at Gestingthorpe down the road, who recently showed me round the collection of ponds he had built on his land last summer. Anna kindly invites me to a series of farm walks she has arranged through FWAG to highlight good practice. She is a whirlwind of warm air. The day is fine. The yellow rattle is growing well in the field. A honeybee drifts around me. It rests on a bramble flower beside me. It is one of Ben and Louise Rees's bees, I can see.

23 June
Ray Davis delivers some muntjac steaks. He sits in the driver's seat of his green Land Rover and explains how they're a 'saddle cut'. He says he won't come in for a cup of tea because he's due over in Boxsted for another free delivery of fresh, organic venison.

'Will you take some meat on the bone, too?' he asks, and digs in his cool box, producing another blue freezer bag.

'I'd love to,' I say.

In the evening, I eat the deer steaks, as Ray suggested, with just a knob of butter and some salt and pepper. They are absolutely delicious.

24 June
Reading Leif Bersweden's *Where the Wild Flowers Grow* I am delighted to find a quote from my dear old friend Ronald Blythe:

> There are a great many ways of holding on to our sanity amidst the vices and follies of the world, though none better than to walk knowledgeably among our native plants.[8]

In the evening, I stand in the field and watch and listen as a song thrush sings from the top of the cedar tree. The repertoire runs for a minute or more. I simply stare, awash in the music, the bird framed in my vision by young oak trees.

25 June

In the morning, I chop up all the vegetables I can find in the fridge and put them in the slow cooker I have borrowed from Maddie along with some vegetable stock and the remaining clumps of muntjac deer Ray Davis gave me. All day long they stew gently, while I go into Colchester with Eva, and come back and go about my ways. In the evening, Maddie and I eat the stew. It is as tasty as you would imagine. In the pot, there is a small scapula bone that has come free of the flesh. For some reason, I want to keep it – as some kind of memento of the deer, but also something more, as some way of showing gratitude for the meal, for the sacrifice of the animal. Then that seems too strange so I simply put it in the compost. Now I think I might have to fish it out again.

27 June

My efforts to bring nature back to the field continue to fill my thoughts and time. Ray Davis has shown a more directly interventional approach to managing a small plot. The depth of birdsong on his five acres clearly seems to show that killing off all the magpies, jays and squirrels has an impact. But I don't feel I can do the same in the field. It isn't in my blood.

Yet I do feel the need to reach further beyond the boundaries of my field, to learn more of larger-scale projects of ecological restoration. I need to know more about rewilding. It is time to hear from some of the experts in the field, so to speak.

Some time ago, I spoke to Guy Shrubsole. That was when Guy was still at Rewilding Britain. He told me he was staying until COP21 and then leaving to write a book on Britain's lost rainforests. Guy spoke of his concern over a dilution of the term 'rewilding'. He suggested that I contact Steve Carver, who is director of the Wildland Research Institute at Leeds University.

'He's the person to ask about rewilding,' Guy said.

Today, when I finally catch up with Steve Carver it is via Zoom. He is preparing to leave for the States for three months to hike the Pacific Coastal Trail, but is happy to meet online. I particularly want to ask him about the whole notion of rewilding. I've already read Steve's paper, 'Guiding Principles for Rewilding', in the journal *Conservation Biology*.

'In that paper, I was just the secretary,' Steve says with a laugh. 'But you probably got the idea that at different geographical scales, rewilding means different things.'

Exactly.

'That paper was trying to bring rewilding back into its ecological roots,' says Steve. 'Rewilding is an ecocentric approach to nature and landscape restoration.'

It is about seeing a basic distinction between rewilding, which is nature-led but human-enabled, and ecological restoration – renaturing – which is human-led and with a distinct desired outcome.

'There's also a distinction between active and passive rewilding,' explains Steve. 'Active involves some human intervention, even if just initially, to get a habitat or ecosystem up to a position from which it can look after itself. Passive would be land abandonment. That often happens by default as people turn their backs on certain lands.'

'There's a fundamental issue there,' he says. 'We can make these small spaces for nature, but nothing exists in a vacuum. Everything is connected. For rewilding to have a genuine impact, it has to happen at scale. That's not to say we need to rewild everywhere. We can't. We need to grow food. We need places to live.'

Steve explains how he's currently working on modelling ecological potential and connectivity across Britain.

'We can then see where in Britain to focus on. Up until now there's been a scattergun approach towards rewilding. Kind of saying, "Anyone got any land they want to rewild?" Which is all well and good but now we need to be more strategic – saying, "We need to target this area here", so as to ensure better coverage.'

It's about creating connecting corridors between areas that are already wild. He's already done the same across the whole of France. The French government is using such information to target areas towards ensuring 10 per cent of the landscape is under strong nature protection.

'Everyone can still do their bit,' Steve says. 'You've got some land and you're doing stuff on it which is great. But it's not really rewilding!'

As he says, it's about scale.

'My bit is not mowing the lawn,' he adds.

I nod.

'Sure,' I say. 'I see the continuity between the window box and not mowing your garden and me with a couple of acres. That feels the same sort of scale. I mean, I'm not releasing beavers into the field . . .'

We both laugh.

Steve asks if I've seen Monty Don's comments about the Rewilding Britain beaver garden at the Chelsea Flower Show, which has just won a gold award.

'He's been questioning whether it's really a garden.'

It's another matter of definitions, a matter of scale. For a little while we can both be amused at the notion of beavers being introduced into gardens across Britain.

Then I start to ask about something that feels rather less funny – that there's a worrying divide within the rewilding of Britain. We're getting to an issue that I've been concerned about for a while. I start to talk about my thoughts on how rewilding seems to be so tied up with other matters – of landownership in Britain, access rights and the capacity for the general public to roam over these isles.

'Maybe it's an access issue,' I say. 'If you start to put big fences up around large areas of Scotland or eastern England or wherever, "because of the beavers or the bison", or whatever, that can have serious implications for common rights. How do the general population of Britain feel they're part of a desire to rewild Britain?'

I can see Steve nodding.

'Yeah, sure,' he says. 'I'm with you 100 per cent. I feel extremely uncomfortable with some of these issues. What's the best way of putting it? . . . It's potentially just another way for the landed classes to keep hold of their land and extract money from the state.'

'Exactly,' I say.

Steve starts to talk about the Knepp Castle Estate in West Sussex — the best-known example in Britain of what most would call rewilding, which Ashley Cooper had mentioned to me just a few weeks ago. The land is owned by Sir Charles Burrell. Back in 2000, he and his wife Isabella Tree turned the 3,500 acres of their farm over to the wild. Since then, they have seen fantastic improvements in biodiversity, with rare species such as turtle doves and purple emperor butterflies being especially newsworthy success stories.

'Everybody thinks Knepp is wonderful,' he declares. 'But what is missed out in the narrative is that it was abandoned land in the first place. It was only some years after they'd stopped farming that Ted Green [ancient-tree legend] and Frans Vera [a renowned ecologist] came along and said, "You need some large herbivores in here".'

Steve pauses.

'I mean, it's still a farm. It's called rewilding but it's still farming because they're extracting a meat harvest — free-range beef and venison and pork.'

He smiles.

'And selling it to expensive restaurants in London with a huge profit margin.'

Steve explains the delicate balancing act required when harvesting free-range meat from a rewilding project like Knepp.

'If you take too many animals off to sell as meat, you're going to reduce the grazing pressure to such an extent that natural scrub regeneration is going to outstrip the grazing. The landscape is going to gradually revert back to canopy woodland. And without natural apex predators, if you don't take *enough* off, then the natural fecundity of the animals will mean that herd numbers increase to such an extent that the grazing pressure outstrips regenerative capacity.'

That is a potential problem.

'You're going to end up with Knepp-vaardersplassen.'

I get the reference – it is to what happened to the pioneering Dutch rewilding project at Oostvaardersplassen, where a small number of herbivores were introduced – thirty-two Heck cattle in 1983, then twenty Konik ponies in 1984 and thirty-seven red deer in 1992.[9] With no carnivores to hunt them and no way of leaving the fenced area, by 2016 the combined population had grown to an unsustainable 5,300 animals. Protests ensued. Oostvaardersplassen was condemned as a site of mass animal cruelty.[10]

'Knepp is a project involving human intervention,' Steve continues. 'I would call it ecological restoration of wood pasture landscape by directing a natural process – which is large herbivore grazing and disturbance through the use of a fence combined with a farming practice of expecting a meat harvest.'

Wow, I think.

'That's got interesting implications then,' I say. 'In terms of what we can call rewilding. If Knepp isn't rewilding, then what is?'

'I've called Knepp "rewilding-lite",' says Steve. 'It's still farming, but it's a wild way of farming. It involves fences. It involves human intervention. There is no carnivore pressure involved. So it's rewilding-lite.'

There are other things about Knepp's version of rewilding that don't sit well with Steve.

'They are steeped in that rural gentry tradition of hunting for fox. They say they only allow trail hunting on their ground. But the opening meet of the Crawley and Horsham hunt is held at Knepp each year. This doesn't sit at all well with me with the rewilding ethic. It's a real problem for them.'

He takes a big sigh.

'There's another problem with these private rewilding initiatives. Unless there's some kind of conservation covenant taken out, they are only at the whim of the current owner. When the land gets passed on or sold, the new owner can do what they like with it.'

'Or if it's no longer profitable,' I add.

'Yeah, quite,' agrees Steve.

'And they may get a lot of eco-tourists at the moment, but what happens when someone else sets up a rewilding project down the road?'

'You end up with a crowded market. There are already a number of estates going down this route of "doing a Knepp",' says Steve.

He pauses before he goes on.

'Obviously, Knepp were in there first. Charlie and Issy have done a really good job. But there are some serious issues we need to grapple with, in terms of this kind of model of rewilding. Our problem in this country is that we don't have large areas of state-owned land – unlike in America. Even most of the national parks here are privately owned.'

Without the capacity to oversee large areas of land for rewilding, there can only be a far less effective piecemeal patterning across the country. In the United States, and other parts of the world, the government can more easily direct conservation initiatives. In Britain, the pattern of landownership is more patchwork, with wealthy individuals who own country estates of a few thousand acres controlling much of what happens on the land.

'I think there's a lot of wave-riding,' adds Steve.

'Amongst that landowning elite?'

Steve nods.

I laugh. It's kind of funny and kind of not. The notion of a model of rewilding whereby landowning estates convert part or all of their lands to managed 'wilderness' which provides an income through grants, tax advantages and eco-tourism is rather different from how most would want to see rewilding in Britain.

Steve tells me of a recently established company.

'They look as if they're setting themselves up to capitalise on exactly this – the country estate rewilding model.'

For a moment, I'm rather stunned by this notion. I'd rung Steve to talk about issues of scale in rewilding. Now, I'm suddenly concerned as to whether rewilding is becoming a way of monetising the natural world.

'I wanted to make this a book about how any individual can be involved with bringing back nature, how everyone can help,' I say. 'But it does seem there are real problems with rewilding in Britain. I mean, you've been looking at this for years.'

'Yeah,' says Steve, with a clear air of frustration.

We chat for a few more minutes. It's obvious that what Steve wants to happen – as the vast majority of people in Britain and across the world surely want – is for a coordinated and coherent approach to rewilding to take place. He has the skills to help map vast areas of land across countries and continents so that rational decisions can be made about the best regions to be targeted. What would really help is if the largest landowners got on board and supported a wider vision of rewilding.

Steve tells me of an organisation called Wild Card who are actively lobbying the Crown Estate and the royal family to rewild their lands. The penny drops.

RENATURING

'Are they the lot that are pushing the Duchy of Cornwall to rewild?' I ask.

They are.

'The Duchy of Cornwall and the Duchy of Lancaster don't want to know,' says Steve. 'The Crown Estate, on the other hand, are interested. We've had some constructive conversations with them.'

Steve explains how Wild Card are seeking to use the mapping he and his team have conducted across Britain to highlight exactly which areas are most valuable for rewilding so as to ensure connected coverage across the lands. Except there's another issue to consider – much of the Crown Estate is also valuable farming land.

'You don't want to be rewilding grade one, two and three agricultural land.'

That crucial connection to farming is where we leave matters. It seems a good place to step away. Steve has a few days to go before he's off to California. He leaves me with a whole mass of questions about rewilding and a flurry of notes.

I sit at the table and try to think.

All I had thought of as rewilding has been thrown into the air. I had understood something of how there had been an erosion of the term as people spoke of rewilding themselves, or rewilding a window box, but to see the best-known rewilding project in Britain as 'rewilding-lite' is startling. The restoration of so much wonderful wildlife at Knepp was fabulous to see, as Steve agreed. I had heard gushing tales from friends who had been on the eco-tourism trails, the safaris, around those thousands of West Sussex acres that now had so much more biodiversity than they did twenty years before. I had marvelled at the stories of white storks that had returned and settled at Knepp. Clearly, the rewilding project there had been amazing at returning life to the depleted soils of the farm. Yet really to achieve serious rewilding across

Britain — as experts like Steve Carver aim for — we need to look at an even bigger picture. We need to have a far more strategic approach using spatial mapping across the country, even aiming to introduce carnivores like wolves and lynx and look for less and less human intervention.

It is all rather overwhelming.

I change back into my field clothes and step out of the back door, weaving my way past the bees and the bramble. In the field, I return to the gentle task of teasing the winter detritus of twigs, leaves and moss from around the young wildflowers emerging on the bank. It is suitably meditative, patient work. I release a frond of yarrow to the light. This is my scale of nature restoration.

On Rewilding

Now a commonly heard word, the term rewilding was actually only first used in print in 1991 in the magazine *Wild Earth*, linked to the Wildlands Project, which aimed to form core wilderness areas across North America. These large-scale landscapes were to be connected by land corridors with the goal of creating wide spaces for apex carnivores to roam free of human activity. This radical, deep ecological approach was driven by a group of US conservationists spearheaded by Dave Foreman and offered the earliest concept of rewilding.

Since then a number of definitions of rewilding have evolved. The main focuses of the term have included the replacement of mega-fauna 'lost' in the Pleistocene, the deliberate reintroduction of specific species into a landscape and the active abandonment of agricultural land accompanied by a passive human management approach. Writers and thinkers on ecology have tended to focus on one or more of these key themes of rewilding in the debate. The environmental charity Rewilding Britain defines rewilding as 'the large-scale restoration of ecosystems to the point where nature is allowed to take care of itself. Rewilding seeks to reinstate natural processes and, where appropriate, missing species – allowing them to shape the landscape and the habitats within.'[11] The rewilding vision offered by the environmental journalist George Monbiot argues more specifically for bringing back large carnivores and herbivores to roam what are now sheep-cultivated farmlands in Britain. In its short history, the word 'rewilding' has certainly garnered a number of definitions, though at its essence it is to do with making the natural world a little wilder (now often expanded into rewilding ourselves, making our human lives a little wilder, too).

In her academic review paper 'Rethinking rewilding', Dolly

Jørgensen saw the term 'rewilding' as a 'plastic word' that has 'tended to conflate several of the discrete scientific uses' as it has shifted to more common parlance. In its continuing evolution as an ecological term, rewilding has rather come to mean what you want it to mean.[12]

29 June

Two days after speaking to Steve Carver, I meet Annie Randall and Hazel Draper from Wild Card. After some initial introductions, I plunge straight in.

'So you're arguing for rewilding of 50 per cent of the UK?'

'That's it,' replies Annie. 'The main focus is that big, single demand: 50 per cent of the UK.'

We're chatting on Zoom. I can see her face on screen, smiling and friendly, and then see her sigh momentarily.

'Which, perhaps, we realise may feel unachievable in a short space of time . . . but the smallest thing it can do is open up the window to other rewilding groups so as to make it more realistic.'

'It fits with the Half-Earth Project,' adds Hazel. 'The idea that we share our earth. So why not actually *share* the earth?'

I'd only recently heard about Wild Card's campaign for 50 per cent rewilding of the United Kingdom.

'For me, the thing that appealed with Wild Card was that directive of going to the lands of the royals, the Church and Oxbridge. That was genius. It certainly felt radical and political. Those are your three central areas of focus, aren't they?'

'Yeah,' agrees Annie. 'Undoubtedly. We've realised it's not just about getting landowners to rewild. It's about a change of approach to the way we manage the land, which is going to be radical because we've got such a romanticised history in the UK of the land and how we relate to it.'

RENATURING

'Farmers have been pushed in a particular direction where yield is everything,' says Hazel. 'But at the same time, many of them are seeing how much harder they're having to work for everything, how the soil is so degraded.'

She tells of that traditional image of the countryside of gulls following the tractor as it ploughs the fields, for the meal of worms unearthed.

'Now there are no gulls following the tractors.'

Hazel has already told me she's an English secondary school teacher who lives near to Hebden Bridge, a market town in West Yorkshire. She has been involved in tackling the ongoing issues of flooding that have been such a problem in that region. She speaks with a calm, clear, steady voice.

'There's this false dichotomy between farming and nature – between food security and nature – that really has to be broken down,' she says. 'Our harvest depends on us rewilding. Our harvest depends on us living with nature. The only way we're going to survive is if we rethink food and farming.'

'Yeah,' I say and nod back to her image on the screen. She's put the matter very well. There's just a short moment where all three of us are quiet, in reflection of that basic truth. In Britain, rewilding and farming are so closely linked.

We talk about how any discussion of rewilding so quickly involves issues of who owns the land and how it is being used. I give the example of grouse shooting, or pheasant shooting down around me in the south-east of England. These activities are tied up with class issues, with wealthy city dwellers working in finance coming to play the upper-class landowner and shoot and drink for the day.

'Yeah,' agrees Annie. 'It's about reshaping the rural economy.'

Hazel tells how she's a volunteer flood warden at Hebden Bridge.

'We live downstream from a grouse moor. They talk about

how much local businesses thrive on the money brought in from grouse shooting. But I saw how badly affected local companies were by flooding which can be directly related to the land being maintained as grouse moor.'

It is the Walshaw Moor Estate that Hazel is talking about. It has been the focus of controversy and environmental concern for a number of years.[13]

'Each year we spend billions on flood defences,' she says. 'It's not only down to grouse shooting. Climate change and poor land management have made things so much worse.'

'There are tensions with people buying swathes of land to offset in order to make money, and they're not really committed to rewilding,' Hazel says. 'So that has been mentioned by people. Often there are issues with not considering the local community, especially when you have tenant farmers losing their land through that process.'

I nod at the screen.

'It's a pity you can't put a law on morality,' she adds and laughs.

There's certainly something in that. The comment reminds me of the phrase used by Anna Beames to describe the unsung local farmers she works with, how they oversee their lands with a 'deep integrity'. That aspect of a moral framework to the way in which people manage the landscape is one that is often assumed, it strikes me. Yet the history of the twentieth century is riddled with actions which have been carried out to maximise financial profit at the expense of the natural world. Back in 1962, Rachel Carson's *Silent Spring* told of the impact of pesticides such as DDT. The collapse in biodiversity, in insect populations, in soil fertility in the last half-century or so across large sections of the developed world only attests to the lack of a far better informed moral framework when decisions are made about farming, wildlife and nature. Rewilding is a way of helping redress some of these matters, but if we aren't careful the movement could be derailed by financial greed.

'It needs to be done right. It needs to be done sensitively,' Hazel continues. 'Not what people are calling it . . . wildwashing.'

It's a new one on me.

'Wildwashing,' I repeat. 'That's a very good term. Like greenwashing.'

Hazel smiles.

'You like that? Shall I claim it?'

There's a lovely breaking of tension. The three of us laugh together.

'Sure,' I say.

'It's probably Steve's phrase,' she says.

For a moment, we can laugh. It feels lovely for me to have met up with Steve Carver, and with Hazel and Annie of Wild Card, who are all dedicated to seeing rewilding implemented in Britain in the best ways.

There is huge variation across wider Europe and the rest of the world in the ways in which land is managed and owned. Annie tells me how she's recently been in Switzerland doing a talk on rewilding.

'You explain how 50 per cent of the land is owned by 1 per cent of the population, and people go "Oh, yeah, you've got a king . . ."'

She smiles and raises her eyebrows. I laugh.

'You start explaining how the queen or king owns all this land and how it's a semi-feudal system and that kind of blows people's minds and they think we're all just peasants. That's the image they get.'

Annie laughs, too, now.

'Like we're still in the Middle Ages,' she adds.

Britain really is so very different to most other countries.

I've recently been entranced by the podcasts of Ken Layne, whose book *Desert Oracle: Volume I* is open beside me, telling tales of the vast desert lands of the south-western United States and Mexico.

'If you look at bigger spaces like the Mojave Desert, it's millions of acres and federally owned,' I say. 'That land was bought so that other imperial powers – Britain, Spain – couldn't take it.'

I'd just read how the Greater Mojave Wilderness consists of 10 million acres of protected wild landscape. The scale is just immense compared to Britain. It's so much easier to rewild when you're dealing with a vast region, all owned by the government. The historical legacy inherited by modern Britain means most of the land is owned by the royal family and a smattering of aristocratic landowners. That issue of landownership and land access is a factor which has to come into any discussion of rewilding, especially when talking about a country such as Britain.

'The Crown Estate owns 615,000 acres,' I say.

'And there's also the foreshore,' says Annie. 'They own two-thirds of England's foreshore. So it's not just a question of rewilding the land – it's a question of rewilding the water. Any mineral extraction or trawling, they have to approve the licence. There's a lot of profit that King Charles makes from simply exploiting the seabed.'

We turn away from the complexity of Britain's landownership.

'Europe has amazing potential for rewilding. There you can have enormous wildlife corridors,' says Annie.

She tells of a fascinating project that is beginning in Italy and Greece where bears are being reintroduced and there is an active education programme to work with communities on how to cohabit with these new arrivals.

'Rather than saying, "Oh, the bears are going to be here and your dogs are going to be eaten", they're going to be honest and say, "Yes, it might be difficult but let's learn how to live with these new wild animals".'

There's something really fundamental here. That notion of reintroducing apex predators is so key to genuine, proper, large-scale rewilding and yet we as humans often have a reluctance to see it happen.

'That's about us rewilding our own consciousness,' says Annie. 'Understanding that we can't go on being the only apex predators. That's what's ruining everything.'

With Britain, the complex matrix of landownership coupled with the fact of being an archipelago of islands means mapping the country is vital. Overseeing what areas are most valuable for rewilding enables the creation of connecting corridors of wildlife. That takes us back to the work Wild Card are doing with Steve Carver.

'With Steve's mapping, we can go to large landowners and show on paper how wildlife corridors can be created,' says Hazel.

The map he is creating with Wild Card is going to be multilayered and interactive so that people across the UK can look at where they are, who the local landowners are, and the rewilding potential.

'It's known as the "Rewilding Mega-map",' says Annie.

Such is the wonder of the coordinated approach to overseeing the rewilding of Britain that Steve Carver's spatial modelling will provide. Then there can be a genuine assessment of what areas are best for rewilding, or native woodland, or for farming. Rather than the piecemeal approach currently operating, with individual landowners dictating where rewilding happens, a more comprehensive and complete programme can operate – with Wild Card leading the pressure on the Crown Estate, the Church and Oxbridge as the leading landowning organisations in Britain to fully endorse and support the way forward.

It does feel good to know that there are people so dedicated and serious in the endeavour to rewild Britain. Annie, Hazel and I say goodbyes for the moment, waving to our screens from across the country. I smile. It's nice to remember that there are many good people out there striving to do good, seeking to make the world better, wilder.

On the Politics of Rewilding and Wildwashing

Rewilding at its core is about letting nature take control, putting it back in the driving seat, allowing it to find its own way – on a big scale. Rewilding is essentially not about making money but about natural restoration of landscapes that have become depleted, most likely due to the actions of humans. There may well be ways in which a rewilding project can provide an income stream – through eco-tourism, for example – but the driving force of rewilding is an environmental one.

And yet we live in a capitalist system. Many who own land seek to maximise financial profit from that land – farming is one way, so is shooting grouse or deer, or building a golf course, or even offering fishing, glamping and camping. In a land like the UK, where much of the ground is owned by a small group of extremely wealthy landowners, there immediately arises a problem around rewilding. How to persuade them to undertake ecological restoration of their lands? In order to operate on the large scale that rewilding really needs, these landed few who actually own the wilds of Britain and Ireland need to be on board. The trouble is that many aren't willing to return their lands to nature, and so rewilding across Britain becomes patchwork and piecemeal. Contrast that with parts of the US, where federal ownership of vast areas of land has allowed huge rewilding projects, such as at Yellowstone Park with its 2.2 million acres. The vision of the Yellowstone to Yukon (Y2Y) conservation initiative takes rewilding way further, to form a project over 500,000 square miles in area – twice the size of Texas, or five times the size of Britain. Creating such a huge area for wildlife is only possible because the land isn't owned by individual profit-seeking owners.

Rewilding is occasionally positioned as a radical reimagining of the landscape and our relationship with nature. In Britain,

George Monbiot's book *Feral: Rewilding the Land, Sea and Human Life* (2013) details how sheep husbandry has been responsible for the depleted state of upland areas of England or Wales (along with grouse shooting and deer stalking). Many reacted against him – outraged at the vision of, say, the Lake District not with shorn green hillsides but woodland and forests. In the US, Peter Michael Bauer's (aka Urban Scout) *Rewild or Die: Revolution and Renaissance at the End of Civilisation* (2008) calls for humans to return to a hunter-gatherer-gardener lifestyle away from our current civilisation that domesticates us, cages us in cities separated from the natural world. Yet for most, rewilding remains a form of natural regeneration *within* the capitalist system, not an argument for revolution.

In Britain, the evolution of rewilding has been dictated up until now by a handful of landowners who have chosen to turn their lands to natural regeneration projects – either as a moral and ethical choice, or because rewilding has offered a way to improve their profits, or perhaps a bit of both. At the Knepp Estate, owner Charlie Burrell turned from farming to rewilding in the year 2000 once he realised 'the clay had won'. The poor state of the soil and an overdraft of £1.5 million were the push factors. The pull factor was a desire to restore nature and improve biodiversity on his land. Now his family's rewilding 'business' turns over £1 million, with a 20 per cent profit margin (compared to the 1 per cent from farming).[14]

So rewilding can be profitable. The example of Knepp has shown other aristocratic landowners in Britain that restoration of the natural world can be a way both of making money and doing ecological good. In England, some 6,000 people own half of the land, so this message only needs to reach some of the privileged few to have an impact on rewilding practices across the country. But this business-based approach to rewilding leads us to the potential problem of wildwashing.

Greenwashing is where businesses deliberately play on green marketing to enhance their environmental credentials to consumers. Being seen as 'green' has become more and more important in recent times – even to big business, and even if that business is highly polluting. Infamous examples include the VW Golf TDI being advertised as using 'clean diesel', or the Airbus A380 aeroplane that was sold as a 'better environment inside and out'. BP rebranded itself in 2000 from being 'British Petroleum' to 'Beyond Petroleum', and changed its logo to a green flower.

Wildwashing is an emergent cousin of greenwashing where rewilding projects or natural restoration initiatives are in fact driven first and foremost by a thirst for profit rather than through genuine environmental desires. In the UK, as rewilding projects are seen to be commercially viable ways of turning around failing farms, so others may seek to profit by mimicking the model. In some ways, this may seem good – for biodiversity, for natural restoration, etc. – yet an overarching, coordinated and large-scale rewilding campaign across the entire country is the real way to rewild Britain, driven entirely by where the optimal sites are for nature, not where capital can best be made.

30 June
Six-spot burnet moth feeding on the marsh thistle. A first sighting for the field.

1 July
Marbled white butterfly resting open-winged among the tall grasses of the field in the early morning sunshine. I could have wept.

RENATURING

2 July
Close to where that marbled white was yesterday, I notice an unusual gathering of plants whose leaves look a little like yellow rattle without the ragged edge. Later, in the study, I look them up. The LeafSnap app says it is marsh woundwort.

6 July
A day beyond the field boundaries, out into the wider world to visit my friend Richard Mabey, the celebrated nature writer. He takes me on a tour of his meadow.

'False oat grass,' Richard says. 'The tall one.'

We are peering down into the mosaic of the meadow, having stepped from the inside to the out.

'Look at your yellow rattle,' I say.

In the mesh of greenery, there are a mass of tall shards of yellow-flowered, sawtoothed leaves.

'This is what I'm after,' I continue. 'I've got patches, but . . .'

The yellow rattle here was introduced long before I sowed it in my own field. Now it is everywhere. Eating away at those grass roots – keeping a lid on their expansion.

Richard wrote to me earlier in the year, excitedly stating that he had identified thirty-five species of wildflower then out and flowering in his meadows. It is less now. Those heady days of spring are already behind us. He is a wonderful guide to the nature of the meadow seasons.

'There are three really,' he explains. 'There's the spring lot, which is cowslips, ground ivy, germander speedwell and buttercups. Then you get the ox-eye daisy phase with bird's foot trefoil, and then the late stage, with the knapweed and pink musk mallow.'

He points to a perfect example a few feet away from us. I explain how I've grown a few plugs of this plant to put into the field. Here, there has been no need for such labours.

'The musk mallow has come in of its own accord,' says Richard. 'It's rather nice.'

I half close my eyes. The colour scheme of the meadow is a weave of mauves and whites. If I'd been here a month and a half ago, it would have been yellow and white.

We head on.

'I'll show you some other things as we go round,' offers Richard.

I am here to learn the ways of the meadow. There is so much to know, yet I am fortunate to have such a teacher. Richard Mabey is a legend. His books range from a glorious biography of Gilbert White to the searingly honest depiction of his own depression and recovery, *Nature Cure*. His early books, like *Food For Free* and *The Unofficial Countryside*, changed the way a generation interacted with the wildlife around them. His encyclopedic triumph *Flora Britannica* is a bible for all students of plant life. My copy always sits at the side of my desk.

There are so many aspects of these gardens that I want to know and to harvest for my own endeavours at creating wildflower meadows in the field. The patch he and his partner Polly oversee is sizeable – around an acre of meadow with another acre of cultivated ground, including what they call the Mediterranean garden, where the more exotic species are collected together.

'We'll cut it probably in about four weeks,' says Richard.

But he has plans.

'This year, I might cut it in two tranches,' he explains. 'And leave some to cut much later in the winter so there are different heights of vegetation for the insects. As another initiative, I'm going to leave this section entirely over the winter – not cut it at all.'

'Just to see what happens?'

'Yup,' agrees Richard.

Experimentation such as this is all part of the effective supervision of the meadow. See what happens. Observe the natural world. Even for a botanical expert like Richard, there is a sense of wanting to learn more, a healthy desire to sit back and peer with wonder on the ways in which these patches of nature will evolve. He oversees the meadow.

But I want to know the practical, the nitty-gritty of meadow management.

'So do you use a ride-on mower?'

'We have a very good gardener who's been with us for ten years,' explains Richard. 'He uses a very, very powerful strimmer and then rakes afterwards.'

Rather noisier than a scythe, I imagine. But clearly also effective.

Richard's attention is torn. There is another treasure – a spike of small yellow flowers.

'Agrimony.'

The delights of the wildflowers around us are too enticing. He points to a carpet of even smaller yellow flowers.

'Lady's bedstraw.'

We arrive at the latest feature in the meadow – a new pond.

'It's been sensational,' declares Richard. 'We've had three species of damselfly, two species of dragonfly, within weeks of it being full of water. And lots of odd boatmen-like creatures that I can't identify. I've had to send pictures of them off to Bob Gibbons.'

'Even you have people you send photos to,' I say.

'Oh, God yes,' says Richard.

The simple act of digging a pond brings so much life, so quickly.

'It's amazing how the pond skaters get on there within a day,' I say.

'I know,' agrees Richard.

'I've never known how they manage that.'

As we head over to the copse of trees, Richard explains how the meadow evolved. He practised what he calls 'adventitious sowing'.

'Whenever there was a bare patch – because of moles, or the tractor mower had scalped it – I'd hurl a handful of stuff in there, so it was very arbitrary, opportunist sowing rather than anything planned.'

There were other ways in which Richard had helped these lands to become the wildflower haven they now are.

'I set fire to a lot of it,' he says.

'Little sections at a time?'

'Yeah,' he laughs. 'That was fun. It was very unscientific.'

I tell him of my exploits cutting back the bramble to create space and then burning great pyres of the stuff.

We've reached the trees. Richard has found something intriguing in the ground beside us.

'It just amuses me how different plants get around in this garden,' he says and laughs again.

At our feet is a tall clump of green, arrow-shaped leaves topped with fists of yellow flowers.

'That's . . .' he begins. 'Erm. What the hell is it?'

I don't have a clue.

'*Phlomis russeliana*,' Richard declares. 'From what was once called Abyssinia. It found its way somehow from the Mediterranean garden down there to up here. And doesn't it look great?! Under the oak and beech trees.'

We both laugh.

'It looks like it's been there for years,' I joke. 'Settled in nicely.'

Yet there's a lingering, more serious question that I have to ask Richard.

'But that's a non-native,' I say. 'You're not too worried about it?'

'Not in the slightest,' says Richard with another laugh. 'I'm delighted to have it here!'

We both laugh again.

'Because there would be some hard-core gardeners who would pull it up,' I say.

'Oh, yeah,' Richard says.

He tells of hearing someone on the radio some years back; listening on a bright, spring day – 'blue skies, swifts around' – discussing the problems of Spanish bluebells mixing with British ones.

'He said, "If I come across any of those Spanish, I trample on them . . .",' Richard says, imitating the voice.

I laugh.

'And I thought, "My God, mate".'

'It's a narrative of invasion,' I say.

'Absolutely.'

I put on a thuggish accent. 'I don't mean to be racist but . . . those Spanish bluebells.'

'Exactly,' adds Richard adopting the same voice. 'Those *Dago* bluebells.'

We both laugh.

'He didn't quite say that, but it was the viciousness of his tone. If he'd said it in moderated terms – "I try to get rid of them" – it would have been very different. But he wanted to go in like a boot boy.'

Our progress is halted by the appearance of a hedgehog sat perfectly still on the pathway before us.

'Ah,' I say.

'Not sure you should be out at this time,' Richard says, peering in.

Just as we are both starting to get concerned, the hedgehog raises its head and trots off into the undergrowth.

One of the latest 'features' of the garden Richard is keen to show is a thicket he has been seeking to encourage that has formed from an oak branch which has split and brought down a hazel and an elm, crashing into a Spanish chestnut.

'Looks like you could make a good den in there,' I say.

'Hidden in there are hawthorn seedlings, maple seedlings . . .' Richard says. 'It'll be fascinating to see how they work it out.

There's that great phrase from John Ruskin, "little quarrels in the family", about how oak branches sort themselves out.'

I tell of an oak branch in the field that's leaning over rather dramatically which I was tempted to chop, then decided it was best to leave it.

'It can fall off or do what it wants,' I declare.

'The narratives they then unfold are wonderfully unpredictable.'

A blackcap's song breaks in.

Before us is a patchwork of primroses.

'There's a fabulous symbiotic relationship between primulas and ants,' Richard says. 'The main agency for spreading primroses and cowslips is ants. The seeds have a tiny bag of fat attached to them. The ants use that – eat it, take it back to their grubs, and while doing that they carry the seeds around.'

It's a glorious example of how complex the processes of wildflower generation can be.

'That's the reason you see primroses colonise, vertically moving up a bank. It's down to the ants.'

I think of my field. This is exactly what has been happening on the south bank, where tiny sprays of new primroses have emerged over the years – clearly thanks to the subterranean endeavours of an unseen population of ants.

Our wander is happily interrupted by the appearance of Polly, who has returned from the exercise class she runs for more elderly local folk down the road at Diss. Richard tells of the hedgehog we bumped into. Polly is thrilled.

'Took off like an express train,' he says, and we all laugh.

Polly joins us on our meadow wander. I begin to tell her of my endeavours in the field.

'I've been digging up as much of it as I can to eventually create something like this,' I say. 'At the moment, it's a mishmash of cornfield annuals. They call it a nurse that you put in with the wildflower seeds to give some colour in the first year.'

RENATURING

'Poppies and cornflowers are incredibly pretty, aren't they?' says Polly.

'That's right,' I say. 'And corncrake . . .'

There's a slight hesitation.

'Oh yes,' says Polly politely.

In the momentary silence, I notice my slip.

'Corncockle!' I shout. 'Not corncrake!'

We laugh.

'I was reading how farmers used to try to get rid of corncockle as it was toxic and would get into the bread and affect people badly. The seeds are big, so they're fairly easy to remove.'

'I'm trying to remember what it was used for,' says Polly. 'Was it a vermifuge?'

'Oh, gosh,' says Richard. 'Might have been. It's certainly quite poisonous as seed.'

'It also has some kind of health benefit,' says Polly. 'I remember putting that in my dissertation. I should go and look it up! We have had some here. I think I nicked some from Norwich Cathedral herb garden.'

Polly was instrumental in setting up that garden. She wrote her degree dissertation on the healing herbs of East Anglia. Along with a friend, she had persuaded the church authorities to hand over a portion of the land around Norwich Cathedral for the creation of a herb garden.

'I got intrigued by that old monk St Benedict, who was all about gardening and health,' she explains. 'Then there was Hildegard of Bingen and all that crowd.'

Her attention is drawn by a flower beside us.

'What have we got here, darling?' she says. 'It's something oniony.'

As Richard recalls the name, Polly is already nibbling at the leaves. It's a lovely moment of insight into their domestic life. There's a delightful intimacy such that you can imagine them pottering about the gardens and the meadow together, doing just this, chatting and musing on the nature of the plants around them.

'It's *Allium neapolitanum*,' states Richard.

'Have a taste of that,' says Polly. 'Seriously oniony.'

'You really can smell it,' I add.

'Think I planted that years and years ago,' says Richard.

We wander on. I want to know what happens to the cuttings from their meadow.

'First couple of years, we put them on Freecycle,' says Richard. 'People came and took armfuls for their guinea pigs.'

'I might have to come back,' I say quietly.

'Come with bags and bags,' says Polly.

'If you leave it until August to cut, most of the seeds are shed,' says Richard.

He says he will wait and see how things go before deciding when to cut.

'Everything is slightly late,' he states. 'Yarrow isn't out at all yet. Wild carrot isn't out in the meadow, though it is in the front of our garden.'

'So you'll just base it on your knowledge of the flowers before cutting?'

'Yeah,' Richard says. 'For anyone wanting to start a meadow patch, I'd love to hand over some free hay.'

It's a very kind offer and one that I will seek to ensure happens. I tell of Yalda's gift of bee orchid seed.

'We've had bee orchids twice. They last two or three years then they vanish for a few years, then come back in a slightly different place. That's how bee orchids behave.'

I say I'll keep my fingers crossed.

9 July

Chris and Jude Gibson come to visit for coffee and a look around the field. The plan is that they will be able to identify some of the wildlife that has started to appear – both animal and vegetable.

Before we have even got over the green way, we halt to peer at a plant.

'I thought this was a dead nettle,' I say.

'It's a relative of the dead nettle,' says Chris kindly. 'It's hedge woundwort. It's one of the things I always talk about when taking people around. First of all, a plant called a wort is a plant that has worth. Those large, soft leaves were used for knitting wounds. Any plant called something wort, you've got to look for its uses.'

Hedge Woundwort

Jude calls us over.

'There's a nice little moth here,' she says. 'It's a free-range moth, I call it. Not caught in a trap.'

I laugh.

'Not an apple pie, is it?' asks Jude.

Apparently, it's her name for a light brown apple moth.

My eyes finally tune into the tiny creature.

'No, that's *Olethreutes lacunana*, or if you prefer the newer name, *Celypha lacunana* . . .' says Chris with virtually no hesitation at all.

'Mothman,' I say.

'Mothman,' repeats Jude.

This is Chris's nickname to his followers on Twitter. Got a moth you don't know? Post a photo to him and he'll tell you in the beat of a wing.

Our three heads lean together. Before us is a very small moth.

'See, it's got the band,' he says. 'There is a depression in that dark-coloured band which is the lacuna of the *lacunana*.'

'What a lovely word,' I say. 'It's tiny, though.'

'One of the 1,500 species of micro-moths in Britain,' replies Chris.

It has been a while since Chris visited the field. I try to remember the state of the site then. The secondary growth of blackthorn shrubs is now cut and laid down as a twenty-yard-long hump like an ancient barrow. It is flourishing as an emergent copse – a den of wrens.

'This is the original meadow patch,' I explain, pointing to the section before us. Removing the blackthorn has really opened out the space, brought in so much more light.

Chris is distracted by another flying creature.

'Ringlet,' he notes.

'The yellow rattle is doing so well now,' I say.

'Excellent,' says Chris. 'And look how it's suppressing the grasses. Just as it's supposed to.'

RENATURING

'It's great, isn't it?' I say with glee.

There before us is a roughly circular patch of the yellow flower, a somewhat hollowed-out section a couple of metres across that looks as though the grass has been selectively snipped out.

'Which is lucky,' says Chris, 'because it doesn't always suppress. It depends on where the yellow rattle came from, what species of plant it was using as its preferred host.'

He starts to tell how if the seeds originally came from an area in need of nitrogen, such as a chalk downland, then quite probably that genetic form of the rattle would prey upon legumes rather than grass.

'Whereas of course here, the land is massively nitrogen-rich,' I add, catching the point, 'so the rattle is doing a different thing.'

'Exactly,' says Chris. 'It's like the effect of sheep grazing. If they eat the dominant species, they increase diversity; if they eat the non-dominant species, they decrease diversity.'

I peer at the patch of field before us with fresh eyes.

'Look at the ox-eye daisies,' I say. 'They've increased massively this year because they've now got more space to grow into.'

The three of us head a little further into the meadow, stepping carefully among the grasses. I start to tell Jude of my plans to mow this area with the scythe come autumn.

Chris has spotted something.

'Common centaury,' he says. 'Very pale.'

'Ah,' I say. 'That was there last year, too!'

The tiny spray of pink flowers is a delight. Now I look more carefully in the grasses around me, I can see other splashes of that same colour. Like the rattle and the ox-eye daisies, the centaury also seems to be spreading.

Jude has found something else.

'It's a froghopper,' she says.

'*Philanthus spumarius*,' adds Chris. 'The common froghopper.'

They work so well together.

'Oh, we've got a bug,' says Chris.

'One for you, Jude,' I say.

She steps in. She has a pair of binoculars specifically for close work. They, like her, are myopic – specialised at peering into the worlds of the smallest creatures. Chris has a camera. He takes a photo and then zooms in on the image. Together, they chat about the possibilities of what this bug might be. They can identify it later from the close-up photograph.

There are bright-yellow bunches of ragwort growing all over the place yet there haven't been so many cinnabar moths this year. They were affected by the wet period in May. We pause by a clump of slender thistles.

'It's just fantastic for the bees,' I begin, and remember that I sent Chris a photo a couple of months back. He had confirmed a sighting of a gypsy cuckoo bumblebee.

Jude has found some young cinnabar moth caterpillars, crawling in the crease of a ragwort leaf, the black and yellow patterning visible even on their pencil-line thin bodies.

'Pila, pila,' she says. 'That's what our granddaughter calls them.'

There are actually quite a few tiny cinnabar 'pila' tucked away about us. Jude has also spotted a couple of mating hoverflies.

'Spira . . .' she starts, 'Spiracoria, or whatever they're called . . . Chris.'

'*Sphaerophoria scripta*,' he says.

'Something or other,' says Jude.

I start to tell how I seeded the patch beside us with both meadow seed and a 'nurse' of cornfield annuals.

'I was told early on that you need patience to develop a meadow,' I say.

'And it will get less flowery over time,' says Chris. 'This is why I'm so keen for people not using the term meadow for the plantations of annual plants. It devalues the whole concept of meadows.'

RENATURING

They had recently visited Cambridge Botanical Gardens, where cornfield annuals had offered a fine show.

'Swathes of them,' says Chris. 'The Corn Quintet: five species that are all called corn.'

He starts to name them.

'Marigold, Flower, Chamomile, erm . . .'

'Cockle,' I add.

'Yeah,' says Chris. 'And Poppy. They had them growing up to here.' He holds his hand up to his waist.

Common Poppy

I'd also seen the display.

'The bees were buzzing, particularly where they had some phacelia in there, too,' Chris adds.

He is caught by a passing moth.

'Silver Y,' he says.

The field is alive with insects.

'There are so many marmalade hoverflies out at the moment,' I say.

Jude has spotted something else. She calls us over.

'That's a speckled bush cricket,' says Chris.

'Not a grasshopper?' I ask.

'No,' Chris says. 'Because it's got the long antennae. Related. But not actually a grasshopper.'

The sun starts to appear from the cloud. We begin to talk about how busy it's been with insect life on the bramble.

'Here's another one,' I say, pointing to an iridescent green bug.

'Thick-thighed beetle,' states Chris.

I laugh and peer closer.

'Oh, yeah.'

The creature really does have thick thighs.

'That means it's a male,' says Chris. 'The ladies don't have thick thighs.'

It's great to have Chris and Jude carry out this informal bioblitz on the field. There is simply so much life that I wouldn't even notice.

There's something else here, too.

'Crab spider,' says Chris. '*Philodromus*? In fact, that might be *Xysticus*.' He calls over Jude to confirm. As she gets her binoculars out, the creature crawls underneath the leaf of the ragwort.

'Probably *Xysticus*,' she agrees, her head upside down, trying to get a better look.

The marmalade hoverflies are everywhere. Chris tells how they've been really worried about them this year but then suddenly over the last couple of days they've appeared in numbers.

'Whether it's emergence or immigration, we don't know,' he adds. 'They do migrate in large flocks.'

'They like the chamomile, don't they?'

Jude has spotted another creature.

'Hogweed bonking beetle,' she says. 'But he's not bonking anyone at the moment.'

I laugh.

'He's not on hogweed either,' adds Chris.

'No, but isn't he lovely,' says Jude. 'They look fierce, being so red, but he's harmless.'

We pause by a patch of cornfield annual flowers that have grown especially high.

'This is where I burnt a load of bramble,' I say.

Chris explains how it's a combination of nitrogen and phosphorus that has created that extra nutrient push.

Jude spots a cinnabar moth flying over the grasses.

'Oh good,' I say.

I'm keen to get Chris's thoughts on what's the best thing to do for improving the meadow.

'Richard told me he'd seen thirty-five different wildflowers in his meadow,' I say.

Chris is an old friend of Richard's.

'He was saying to just treat it roughly,' I say. 'Cut it all back, burn it . . .'

Chris explains about the new meadow area over at Beth Chatto's gardens, where he acts as a wildlife consultant.

'That's come up without any seed at all and it's looking superb.'

He suggests the best thing I can do is wait.

It's such an important message. Let the natural landscape grow of its own accord.

Beside us is another tiny wonder, which I've just noticed.

'Scarlet pimpernel,' declares Chris. 'They seek him here . . .'

'Isn't that lovely?'

'It will disappear when the meadowland closes up. It's a ruderal species. What you'd call a weed. It follows humans in their wake because we disturb ground. Its aboriginal landscape is naturally disturbed places like sand dunes or mobile screes. It's a camp follower.'

'Like poppies,' I say.

'Yes, exactly. Its seeds can remain dormant for long periods of time – again like poppies,' continues Chris. 'I suppose originally they were adapted to disturbance by aurochs and such like . . . if that only happens every fifty or five hundred years, they will just wait.'

We step a little further through the grasses. There before us is an area some ten yards by five, roughly dug down to a depth of a couple of feet.

'The pond,' I announce. 'This is the latest project. Sadly, delayed due to bad knees.'

'Ah, the pond which isn't yet,' says Chris.

'I want to do it by hand,' I say. 'I just don't want to get a digger in and cause a lot of disturbance. Maybe I'll just do it gradually.'

As ever, Chris offers a useful take on the matter.

'Mmm, I mean, yes, it would cause disturbance,' he agrees. 'But bringing a digger in would only be doing what a rampaging troupe of aurochs would be doing 10,000 years ago.'

It's a fabulous image.

'That could sell it for me!' I say.

'There really is no substitute for disturbance,' Chris explains. 'Biodiversity thrives on disturbance.'

He's distracted by a plant tucked away in the grasses.

'Is that more centaury?' I ask.

The tiny pink flowers peer from the green.

'It is,' says Chris. 'Interesting, because that is very much a plant of sandy, gravelly soils. You're on clay.'

'Yeah,' I agree. 'Solid clay.'

Jude has discovered the shallow hole I've dug.

'That's where you're going to have your pond?' she asks.

'One day,' I say and repeat Chris' argument about a digger being like a herd of aurochs.

'Disturbance is natural,' he says. 'It's exactly why coppicing is so good for woodland. Actually, don't take too much care with the digger. It's a big enough area for you to trundle it over. Provided you're not running it over a special orchid or something like that. Then you can watch what comes into the gaps. Biodiversity colonises gaps.'

It's a great line.

'Biodiversity colonises gaps,' I say, repeating his words.

I smile. My original plan had been to build a decent-sized pond, five feet deep.

'Once I started digging, it soon became clear I would need a digger,' I say.

'Or lots of people,' says Jude.

I laugh.

'Well that was my other plan. But no one wanted to do it!'

Now Chris has given me the reassurance to continue the project.

'You spoke of helping to create sectors,' I say. 'Even in a two-acre plot. I remember you saying how it's good to cut down the thick grass in some patches so as to allow space for lower grass species to flourish as well.'

'That's right,' assures Chris.

'I guess it's about being conscious of what you're doing,' I say.

Jude compares the process to what she calls the 'foraging craze'.

'Leaving some, taking some,' she says. 'In a way it's good, with people getting to know nature, but on the other hand you don't want to take everything. Then there will be nothing left for the wildlife that need it.'

They had recently had someone on one of their nature walks who wanted to be helped to forage as they went.

'She said, "Can you take me out and then do cooking with what we pick?"' says Jude.

'She seemed to take umbrage when we said it's not really what we do.'

I tell how I found a giant puff ball down in the woods nearby and had brought it back to the field.

'As a spore source,' I explain. 'It was kicked around the field all through the autumn. Hopefully, something will happen.'

I picture pure-white, foot-high domes appearing.

Stood in the far corner of field, we pause by a patch of thistle. The main, central expanse of the field stretches before us – an acre and more. There are yellow sprays of ragwort evenly spaced within the longer grasses.

'So what should I do with this?' I ask. 'Do I need to cut this all back, or can I leave it?'

'If you leave it, you'll get the very difficult problem of woody growth to deal with in four years' time,' says Chris. 'I would be inclined to cut it back – probably not even on an annual basis – a two- or three-year cycle . . .'

'That sounds better,' I say.

It's a big area to cut down. I have no tractor, after all.

'I have a scythe,' I add.

'It doesn't need to be done every year,' reassures Chris. 'But anything more than three years and the woody growth will take over.'

'Scything and then raking as well. That's important, isn't it?'

'Yeah,' says Chris. 'Keep the nutrients down.'

The future of the field is easy to imagine when stood there – a flourishing, rich body of wildflowers and plants that have been brought in to this space from a variety of local sources. I tell them how Richard Mabey has promised he will bring a bag of hay from his meadow when he visits in a few weeks.

'Brilliant,' says Chris.

Then there is Yalda's present of bee orchid heads.

'The seed is just amazing,' I exclaim. 'Dust.'

But it will all take time. Chris reminds me it will take at least five years before any bee orchids will be seen.

The thistle beside us is an intrigue for Chris.

'That one could be a hybrid. It's got some marsh thistle in it – with that cluster of heads.'

'This is where there were all those cuckoo bees,' I say. 'I wondered if they chose this thistle in particular, if they thought, "Oh, I like that".'

'Yes,' agrees Chris. 'Most bees have preferences.'

A pile of pruning from the apple trees, Chris calls a hibernaculum.

'A bug hotel,' he adds with a smile. 'For insects over winter.'

He has stopped by the walnut tree.

'Ah,' I say. 'This is so lovely, isn't it?'

'What is it?' asks Jude.

'Unfortunately, it's a black walnut,' I say. 'American. I wanted an English walnut, and bought a black walnut by accident.'

'I wouldn't know the difference,' confesses Jude.

'Totally different leaves,' says Chris.

'Do you still get walnuts?' asks Jude.

'Not really,' I say. 'I've had about three.'

That's three in twenty years.

The tree looks magnificent – there is a beautiful symmetry to the frame of the branches, the green crown stretching out thirty feet up into the blue sky. But there are so few walnuts to show for all this growth each year. It is such a shame, both for me and the local squirrels.

'It's a bit like me buying that,' I say, pointing to the tree in the far corner of the field, '. . . as a Lebanese cedar.'

'Mmm,' says Chris.

'I now know it isn't,' I say.

'No, it isn't,' agrees Chris.

'Which was a bit naughty as I did go to a specialist place to buy a Lebanese cedar and was sold that when it was this height,' I say indicating around three feet from the ground with my hand. 'That's what it's turned into.'

Chris nods. 'Well there is a long history of such things,' he says. 'Like James the First . . .'

'The black and the white . . .' I say, laughing.

Jude is none the wiser.

'Brought over the wrong sort of mulberry trees,' Chris adds. 'For the English silk industry.'

We all laugh.

'Someone made some cash on that, don't you think?' I say. 'He was sold the wrong ones.'

'He had to buy them from the continent,' agrees Chris. 'No doubt there was some kind of rivalry at play.'

It is a great story. How an English king bought 10,000 of the wrong mulberry trees – the black rather than the white variety – and the silk moth larvae wouldn't eat the leaves. Or, at least, that is how the story goes. It makes me feel better about my dodgy Lebanese cedar.

Chris is looking at the seed pods on some vetch which Jude has pointed out.

'Common vetch,' he says.

'There's another vetch in there,' I add. 'With tiny flowers.'

'Smooth tare,' says Chris.

He gets closer.

'Actually, this is the hairy tare.'

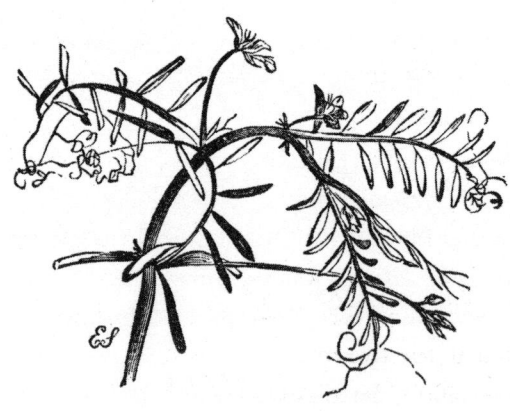

Hairy Tare

RENATURING

Smooth tare has a slightly larger, bluer flower, he explains.

Jude has a baby shield bug in her sights.

'Do you know why they like rhubarb so much?' I ask.

'They'll be dock bugs, probably,' says Chris.

'Oh, so they may be dock bugs,' I say. 'Every time I go to the rhubarb, they're just having a party out there.'

Jude has the baby green shield bug in her hand.

'Second instar?' suggests Chris.

'Yes, probably,' agrees Jude.

There are five instars in total. Three more morphings for this tiny insect to go through – shedding another exoskeleton each time – before it will emerge as an adult form.

She places the creature back where it came from.

'Look at this gorgeous grass,' she says.

'Common bent grass,' says Chris.

I have so much to learn about meadow grasses.

'Common bent grass,' I repeat.

'*Agrostis capillaris.*'

'Compared to this one, which is called Yorkshire fog,' he adds. 'It has to be touched.'

The leaf-shafts of the Yorkshire fog are silky-soft.

We have returned to the southern patch of the field.

'One of the first things I did this spring was walk around and find where all the yellow rattle was so that I wasn't standing on any of them.'

Jude laughs.

There is one plant there on its own before us, away from the main collection.

'They're very close to Chris's heart,' says Jude.

'I studied it for five years as my PhD, and one year as my degree dissertation,' he adds.

He hasn't just worked with yellow rattle; he's a leading expert on it. I know something of how important it is for creating wildflower meadows, but little more.

'So you get different variants of yellow rattle?' I ask.

'Yes,' Chris explains. 'There are two species in Britain. One is this common one. There are five or six genetic types – found in different habitats. So, one in mountains, one halfway up mountains, one in dry areas, one in meadows . . .'

He explains how when he did his research he didn't have access to DNA profiling and that in fact within those types you could probably find genetic races which are adapted to feeding upon particular hosts.

'I think it's not totally phenotypic,' he explains. 'It's genetic.'

Chris gives the comparison to the work which is discovering that there are genetic differences between cuckoos that parasitise meadow pipits' nests and cuckoos which parasitise reed-warblers' nests.

That comparison to cuckoos is so suitable. Each yellow rattle is also a parasite. The plants will attach to host grasses and legumes.

'Whatever suits them best, will make them grow the best,' says Chris. 'So there is a genetic aspect, but there's also a phenotypic angle – what it grows on gives it the biggest kick in life.'

Chris turns his attention to the blackthorn hedge that marks the field boundary to the green lane.

'That is getting beyond its use for birds.'

He explains that what I should do is cut it all down and simply let it regrow.

'For whitethroats . . . or nightingales, if you're lucky,'

The birds like the thicker bottom of the bush. At the moment, the hedge is too leggy to be much good.

So there is another job for me to do. Cut the scraggy blackthorn branches down.

'And can I stash them in there?' I say, pointing to the base of the hedge.

RENATURING

'Yeah,' agrees Chris. 'That creates instant bottom.'

There is so much to do. We arrive back at the entrance way. Our short walk around the field has taken a good hour but the first bioblitz is complete. Chris and Jude have to head off so we say our goodbyes and I head inside to write down a list of the wildflowers that they have identified in the field.

A Wildflower Field List

Bramble
Common centaury
Common field-speedwell
Common knapweed
Common poppy
Corn chamomile
Corn marigold
Corn mint
Corncockle
Cornflower
Cowslip
Creeping buttercup
Creeping thistle
Cut-leaf cranesbill
Fleabane
Garlic mustard (Jack by the hedge)
Hedge woundwort
Herb Robert
Honeysuckle
Ivy
Lady's bedstraw
Meadow vetchling
Ox-eye daisy
Primrose
Ragged robin
Red bartsia
Red clover
Rosebay willowherb
Scarlet pimpernel
Selfheal

RENATURING

Slender thistle
Small blue flowered vetch – hairy tare
Spear thistle
Stitchwort
White dead nettle
Wild rose
Wild teasel
Yarrow
Yellow rattle

Ragged Robin

On Biodiversity

The biodiversity of an environment, ecosystem or patch of land is simply the entire web of living creatures – not only the animals, insects and plants but the fungi, the bacteria, the mesh of micro-organisms. Simply put, the more complex the 'biological diversity' of an area or region, the better. A species-rich, intricate ecosystem such as the Amazon rainforest, which supports some 116,000 species, contrasts to a region such as the Sahara desert, which has a far lower variance of life forms.

In contemporary contexts, the matter of biodiversity is often framed from a perspective of loss or decline. The vast human presence on the planet and the climate changes caused by our actions mean ever-increasing pressures on biodiversity as many species struggle to survive. Cities have a significant dampening effect on biodiversity, known as biotic homogenisation, such that creatures who need specific environments decline and die out while generalists like feral pigeons thrive. Yet climate change solutions and biodiversity are also intricately linked. Half of the carbon emitted by human activity is absorbed by the land and ocean. Different ecosystems have different biodiversity patterns. Some are better than others at sequestering carbon. Forests cover some 30 per cent of the earth and provide over 60 per cent of these 'nature-based solutions', but peatlands store twice as much carbon as forests even though they only cover 3 per cent of the globe. The intricacy of the biodiversity is vital to each ecosystem and so to the capacity of that ecosystem to store carbon. The UN Framework Convention on Climate Change and the UN Convention on Biological Diversity were both established at the 1992 Earth Summit at Rio – agreements by governments to begin to deal with these closely linked global concerns.[15]

Global biodiversity loss means the disappearance of species

from our planet at a rate of up to one per hour and the pace of loss is still accelerating.[16] Biodiversity loss seems a somewhat clinical, scientific term to describe how in the second half of the twentieth century human actions have directly led to the death of millions upon millions of strange and wonderful creatures. The main cause is the huge intensification and industrialisation of farming across the globe. In the UK, the 'State of Nature' report in 2016 detailed the extent of biodiversity loss. Between 1970 and 2013, 56 per cent of UK species declined. Different species declined at different rates. For example, the eyed longhorn beetle had vanished by 90 per cent; the hedgehog by 95 per cent; the lesser spotted woodpecker by 83 per cent. Other species, like the Cullum's bumblebee and the wryneck, have gone extinct.[17]

Renaturing and rewilding projects help increase biodiversity in the garden, field or whatever landmass is under consideration. Striving to help native species by encouraging favourable natural conditions often involves actively working to bring in species not currently present — by wildflower and grass seeds, for example — that then gradually help to create a more complex food chain and ecosystem in that area.

In my field, one of the first tasks was to help support the establishment of various different areas — patches of bramble, meadow flowers, ragwort, ivy — common, vital plants that so many species feed upon. Mainly, that meant one thing — stepping back from mowing and cutting. That simple act of allowing nature the space to exist is the first step to turning the tide on biodiversity loss.

11 July
I sit on the elm log in the borage field. I watch a hare progressing in its own idiosyncratic way, a series of seemingly directionless trots and halts — a jog, back raised, head down. Then stop and

sniff. Head raised. Two black ears rise above the mauve haze of the borage and then off again. For ten minutes or so, I watch this wild creature.

All we need to do is leave the space for nature to live in. If we do not know, do not understand every nuance of every movement of every creature, it does not matter.

Behind me, a crow of some kind – rook, jackdaw, I cannot tell – keeps up a strain of croaks and calls, a chatter that I do not understand. Again, it does not matter to that creature that I am ignorant. It only matters that I and my species do not destroy all the lands, trees, places, where that crow wishes to live.

13 July
It is dawn.

As I step into the field, a buzzard flies down from a perch on high and away south. The sight lights a flicker of awe within me. It feels auspicious.

Renaturing the mind involves learning to see, getting to know the butterflies, the wildflowers, the birds – it is about placing yourself in a space such as this, at a time like this, when a numinous sense exists to the day. Renaturing the mind is about looking, listening: seeking to understand the ways, the patterns, the movements – and perhaps even the meanings – of the birds and the other creatures that share this land.

15 July
There is to be a gathering of Suffolk farmers, discussion of farming practices, then a hog roast. Anna Beames invites me to come along. It sounds like a great chance to step out of the field and to learn more of how others are busy helping nature recover, and especially to hear some of the ways farmers are innovating, bringing in new practices to support the life in the rather larger fields and hedgerows they look after.

Anna sends a photo of the invite. The event is at Framsden Hall, by kind permission of Lord and Lady Tollemache. It sounds posh. So I text Anna to ask her what to wear.

Anna is reassuring. 'It's very informal. All farmers – jeans and T-shirt type of thing.'

The plan is simple: a walk and talk around the fields.

We meet in the grandeur of a fabulously high medieval barn out in the depths of the Suffolk countryside. Anna is already introducing the speakers as I arrive. There are some forty souls standing around listening.

'I was wondering how to encapsulate the Helmingham Estate as a whole,' states Anna. 'Considering it's known for its spectacular historical lineage from Saxon to Norman roots. The barn in which we're standing dates back to 1500, as do some of the original field names – like Tinker Pit, a source of clay production for the roof tiles for this very barn.'

I can't help but look up and around – at the brickwork, the vast arched, supporting beams of oak.

'There is a naturalistic evolution to the emergence of the Helmingham Estate, but most actions were undertaken for a particular purpose, a function,' continues Anna. 'Beauty often derives from function. The original landscape – the reason for the walled gardens – was to protect from marauders and deer invasion.'

The gathered seem like a normal bunch of folks – though they own, or manage, a good percentage of the farmland of Suffolk. I glance about. There's the odd exception – like me. I see my friend Paul a few feet ahead.

'Boundaries. There are many here – it's a complex job unravelling the new from the old. Many pre-date the sixteenth century,' says Anna through the microphone. 'Many other boundaries were destroyed after the Second World War for extended crop production. Field sizes here have ebbed and flowed over the years.'

She speaks with a calm confidence.

'Tonight we're going to go beyond the ha-ha . . .'

A dog barks far off, beyond the walls of the barn.

Anna talks about the notion of 'natural capital' that has come into farming parlance recently. Natural capital, it seems, is about 'carbon calculators, carbon trading and biodiversity net gain'. Each farm has a natural capital value that can be assessed and evaluated. Once ascertained, that figure provides a baseline for each farm – data that can be tracked forward in time to see how the natural capital of those lands has improved (or not).

'As farmers and landowners, you hold a totally unique position,' says Anna. 'The ability to both sequester and reduce carbon emissions – through soil. There is no other sector in the world that can do both.'

It is soil – 'this thin skin that lies between rock and air' – that is the key. Soil is 'the engine room of all ecosystems', and indeed all life on earth.

There is a kind of delightful sense of the preacher to the way Anna speaks. She has a zealous enthusiasm. I've already heard her speak on soil, and heard her enthusiasm for the subject. Now she is letting her flock of farmers see that same passion.

Soil is 'the gut' and there is such complexity within it. We are only starting to understand some of the intricacy of the 'wood wide web', of the mycorrhizal fungi that exist there.

'There are phytochemicals, microbial exchanges, hot-houses of microscopic bacteria,' Anna explains.

I look around at the congregation gathered here in this cathedral-like barn.

'But this communication has been interrupted in our journey towards progress.'

Anna stresses that final word.

'What is progress?' she adds more quietly as an aside, with a smile. 'Perhaps that's a question for another evening. As Roosevelt once said, "a nation that destroys its soils destroys itself".'

There's a slight pause. It's a powerful quote.

'Tonight we're asking you to stretch your vision and your boundaries. This is about long-term security, not short-term gain. It is about sustainability. This is an era defined by rapid change and uncertainty. This is about reducing your vulnerability and risk.'

The environmental crisis is already present. She explains how the capacity to sequester carbon and secure high biodiversity are going to be key to farming in the very near future.

I'm not a farmer but I'm impressed with the manner in which she's directly framing the vision of how farms will need to be reacting to climate changes. I look around. Gathered here are a mere handful of people, but they own and run thousands upon thousands of acres of England's best lands for growing crops, for feeding the population. I wonder how they have reacted to Anna's words. Has she too much of the radical in her call to arms? Farmers are thought of as a conservative crowd by nature. But they have seen the ever drier summers, the increasingly wet winters. They know their yields are reliant on ever more expensive fertiliser. They know the soils are weakened.

'Unstable times require navigation. Today's agriculture is about truly owning your farm, your direction and your decisions. Minimise your risk. Diversify and adapt. Harness natural capital producing healthy food and increasing biodiversity value.'

Anna draws her talk to an end. There's a clatter of handclaps that echo about the brick walls.

We shuffle out of the barn. The next part of the evening is the walk. We are led away to walk the fields, to see some of the best practices being exercised here on the lands of Framsden Hall. Our leader is Glenn Buckingham, farm manager for the entire Helmingham Estate. He's a solid-looking figure with a broad smile who farms these lands with a radical and enlightened approach that is incredibly impressive.

We head out of the barn. I catch up with Paul among the departing crowd. He introduces me to a couple he's already been chatting with – Lucy Manthorpe, who has recently become a farmer, and Neil Gant, who is managing her farm. Apparently, we have a mutual friend.

'So you know Ray Davis, do you?' says Neil.

'Yeah,' I say, and laugh.

'Saw him the other day,' says Neil. 'Told me all about you.'

'Oh, right,' I say. 'Did he bring you some deer?'

'No, I gave him one,' Neil replies.

There's a delightfully dry humour to his voice. I laugh again.

'And he returns it as steaks,' I say.

'He does.'

Neil and I tag along behind the trail of farmers along a track out to the fields of Suffolk. Ray was talking to him about my visit.

'Didn't he try to shoot a jay?'

'I wouldn't let him,' I joke.

'He said,' says Neil with a cheeky smile.

It turns out Lucy has just bought 400 acres of farmland close to Kentwell Hall. Neil is the farm manager/gamekeeper. She and Neil met Ray one day when he simply turned up on the land. Ray had shot there since he was a boy. Now he has become firm friends with both Lucy and Neil.

'He's a lovely man,' Neil says.

We arrive at a barley field and follow a path carved through the crop to a line of newly planted trees, catching up with Lucy and Paul ahead of us. Neil is telling me how he's recently spotted a purple emperor butterfly in some ancient woods he looks after.

'We've only been there since September,' he says. 'It's a new venture. We're due to be having the bat people over, and the moth people.'

Neil is clearly an easy-going country boy. Originally from over at Hitcham, he tells me. He has an engagingly friendly manner. He's also someone handy to have around a farm, being well over six foot two and clearly strong as an aurochs.

The gathered masses settle down.

'Evening all,' a voice pipes up through the portable speaker that Anna is holding.

'I'm a farmer from Cambridgeshire. When we started farming, some of our soils were in a pretty poor state. I wanted to find a way to address that degradation, in a way that made me money – I'm a farmer – and provided biodiversity, plus that met some of the goals around increasing natural capital around the farm.'

He has captured our attention.

'As farmers, we harvest sunlight,' he states.

It's a powerful image, one easily taken on as we stand in a sunlit field with golden barley swaying gently around us.

'Yet there is no doubt that climate change is biting. In my farming career, we've seen the hottest, coldest, driest, wettest summers and winters on record. We need to adapt our farming systems. We have to adapt to climate change. We're also trying to adapt to biodiversity loss. We're trying to improve our economics. We're trying to do a lot of things.'

He urges us all to take a look around the lands about us.

'Lovely,' someone says beside me.

'A fantastic landscape,' the speaker continues. 'It's not a monoculture.'

He pauses. His name is Stephen Briggs, I later learn from Anna, one of the top experts in agro-forestry – the planting of trees within traditional farming crops such as barley and wheat. The result is more productive, profitable land that has improved soil structure and increased biodiversity.

'Nature doesn't do monoculture. Because it's a really, really poor use of resources. We can take some lessons from nature to move away from monoculture.'

He turns to the man beside him.

'What Glenn is trying to do here is to integrate perennials and annuals,' Stephen explains, 'and to farm in a three-dimensional way.'

At that point, the moment is broken. Someone's mobile phone goes off. A woman near me dives into her bag, scrabbling away as Stephen continues to detail his wider vision of farming.

'The barley has started to go golden. It'll be harvested in another four weeks' time.'

Thankfully, the ringing halts.

Stephen explains how any plant, any crop, can be seen as a way of capturing the energy of the sunshine, acting as a solar panel.

'That barley has been photosynthesising for seven, eight months. But we're turning those solar panels off in another four weeks, so it won't be building any more carbon in the soil. At the time of year when there's maximum solar radiation, we actually turn the solar panels off. We're starving the soil.'

It's a fascinating way of seeing standard farming practice.

'The trees around us,' he continues. 'They only started putting on leaf in about April. They'll carry on right through until October, November. So they're using a different period for their photosynthetic capture.'

He pauses.

'Added to which, barley is two feet tall, with a root system about two feet deep. Some of the trees here will have canopies that are maybe thirty feet wide and root systems that are twenty feet deep. They have a different place in the landscape.'

The phrase rings so true. I'm starting to see something of that notion of three-dimensional farming. As much as anything, it seems to be a way of understanding the land and the variety, the complexity, of the plant life growing there – how it all falls under the umbrella of farming when seen in this way.

'The tree roots will be deeper than the crop roots. Importantly, they won't be competing with each other.'

There are nuances to consider. Planting the trees on a north–south axis though a field means they will not be over-shading the crops. Yet the shade offered by trees can also be a plus. For a livestock farmer, a 'sylvo-pastoral system' – ensuring trees are growing across the pasture lands – brings additional benefits alongside increased pasture productivity.

'As climate change bites,' Stephen warns, 'animals will need shade and shelter. There will be more extreme temperatures in the winter and the summer – and wetter periods.'

Sheep and cattle will also munch on leaves that provide different minerals, sugars and starches from those they can get from the ground. Other animals can also do well with this tree-focused farming method, particularly chickens.

'A mobile poultry system under a canopy of trees works really well. It's only us as humans that have taken a jungle fowl out of the jungle and stuck it into sheds or open range.'

Stephen tells how this means that chickens consequently spend much of their time worrying about being predated upon and establishing pecking orders.

'If you can provide an environment where they can integrate with trees or perennial shrubs, that reduces their stress levels. In turn, that makes them more productive, more profitable.'

For many of the farmers around me, that is understandably the key factor. Stephen starts to go into some details about the effects on productivity rates of integrating agro-forestry practice. A few heads around me nod gently.

'This crop of barley only uses 60 per cent of the available solar radiation; 40 per cent is growing weeds – which is why you spray for them. If there was 100 per cent light utilisation by this crop, there would be no weeds. So the question is, how can we capture more of that solar radiation?'

There are new pressures on farming from government on productivity, on biodiversity and on sequestering carbon. Agroforestry can certainly help.

'If we had a 5.8 per cent increase in tree cover, we would meet our climate change obligations from the Paris Accord.'

The statistic is startling.

'It's about the right tree in the right place,' says Stephen.

There's an obvious echo there of Beth Chatto's famous line about the right plant in the right place. It makes so much sense. We need more trees.

Beside me is a wooden stake wrapped in a plastic protective case. I lean over and look down into that inner world. There's a young oak sapling. It's one of a line that stretches north across the field. These are deciduous trees. Stephen tells how he's planted various fruit trees as well in similar lines across his Cambridgeshire fields.

The benefit of planting trees in fields extends way beyond carbon capture, biodiversity and welcome shade. Beneath the surface of the field, there is another vital process being boosted – the mycorrhizal fungal network of the root systems of the trees that so enhances the soil. The exact processes and wonders of these underground meshes have only relatively recently begun to be recognised and understood since the pioneering studies of Canadian forest ecologist Suzanne Simard in the 1990s.

'Mycorrhizas are one of those beneficial fungi in the soil that link everything together,' says Stephen.

He should know – as a student of soil science he was investigating these remarkable growths some thirty years ago.

'The trouble is that as farmers we do everything to try and destroy them. We apply fungicides. We apply ammonium nitrate fertilisers. These have a detrimental effect on these networks in the soils.'

Stephen points a hand down the line of staked saplings.

'These strips of trees create refuges. We're seeing that already at my home farm – ten years down the line.'

He pauses.

'If you can imagine this set-up ten years on, the crops nearer to the trees are yielding higher than those in the middle of the field. They're benefiting from these mycorrhizal associations in the soil that are creating more stable conditions – improved drainage in the winter, better water retention and shade in the summer. We're seeing real ecosystem change.'

Stephen hands over to Glenn Buckingham to explain a few details of these newly planted trees.

'Of these two lines here, the parent oak trees are just over there behind that hedge,' he says, and points over the barley field to a patch of green a few hundred yards away.

'We've collected the acorns and then used their offspring here.'

There are a couple of audible 'ahhs' from the gathering. I look around. There are a few heads nodding approval. But this is a group of farmers who have been around a while. They've seen and heard enough initiatives over the years.

Glenn turns to a matter that might appeal more. He tells of meeting Tony Juniper, chair of Natural England – the official government body for the conservation and restoration of the natural environment in England – and asking him what kind of finance there is behind this new 'natural capital' initiative.

'His immediate answer was £18 billion a year.'

There's a collective intake of breath. A couple of heads nod.

There are questions, queries. Won't there be issues with harvesting the smaller fields? Won't the trees be taking water out of the land? What of the field drains? Won't the roots affect them?

One solid figure beside me with arms crossed over an impressive belly speaks out.

'What happened after t' Second World War? We didn't have the grub, so they said "Pull out the hedges! Pull down the trees!"'

His voice carries the age-old lilt of deep Suffolk.

'Next to where we farm, thar was an eighty-acre wood. That was all burnt. Every tree in it.'

Now we need those trees. We need the hedges. We know now the value of the wild edges to the farmland, the essential goodness that it provides to the soil, the network of mycelia that is so vital to the health of the ground and how without that fungal network crops will not grow. These gathered souls know their lands so well. They know that each year they have to pour more fertilisers, more chemicals onto the fields in order to maintain yields. We all know that this is not sustainable.

I stand in a field in Suffolk with a fine bunch of farmers and we are shown some of the ways of the future of farming. Yet, they are really the old ways – caring for the soils, allowing space for the natural world. What we are being shown are ways of renaturing the practices of modern agriculture.

18 July
Back in the field.

On the day that will become the hottest recorded in British history, I am out early collecting seed. The sun has not yet reached this corner of the field. I tiptoe into the meadow. The yellow rattle has dried now to brown-pocketed stems. Each pod contains half a dozen seeds – like thinly sliced circular slivers of parcel paper half a centimetre across, each with a smudged darker centre and all held by their pod as by a cupped hand, part open to the sky. The pods jangle, rattle to the touch. I lift one stem, turn it up and shake it against the side of the bucket. The seeds fall from the pods, scattering into the wide-open darkness of the bucket, gathering in the bottom.

For fifteen minutes and more I move about the meadow collecting seed. The mass grows in the bucket. The ox-eye daisy seeds are rather easier to gather. The worn, round seed heads merely crumble in my fingers – the seeds are minute pale arrows a millimetre across, five millimetres long that fall from each flower into the bucket in a short shower.

I step back into the cool of the cottage, a still sanctuary from the heat. Now for the next part of the seed ceremony – laying a white sheet over the kitchen table. I pour out the contents of the bucket to form a rustic pyre of plant smatterings, mainly yellow rattle seed – a brown sea of slivers, a mess of tiny circles that seems to shimmer, to move. That movement is due to the actions of the various creatures caught in this process, insects rudely shaken from their homes in the field.

There are a number of small black beetles with an incredible jumping capacity that leap around the white sheet and onto the kitchen floor. Then, there are what look to me like shield bugs that are larger – a centimetre across – and emerge from the mass of seed more slowly. A long-horned beetle of some sort steps from the brown jumble with calm, languorous intent. I stare into the morass and see there are other minute creatures, tiny things a millimetre or two in length. Some sprint to escape, others tuck and hide as my Gulliver-like fingers brush the pile, seeking to separate the detritus of seed pods, leaves and living beings from the seed. I do try to save them – catching them in a jar to be returned back to the world they have been so suddenly torn from, but I know that some will not make it, that some will remain in this mass to be suffocated and starved as they are accidentally gathered up with the seed in a big brown bag to be stored until autumn.

I think of the Jain of India – a religious order whose followers go to extreme efforts to ensure they do no harm to any living creature. I see an image of a figure brushing the dry, dusty ground before them with a plume of feathers as they walk to guarantee no insects are harmed by their feet.

For a few more minutes, I leave these creatures to make their own way to safety. Some edge into the pale white desert beyond the pile of brown seed. They will be saved, lifted by a benevolent God and returned miraculously, Jonah-like, to the field.

A scorpion-like spider, perhaps four millimetres from pincer to pincer, shrinks – arms pulled close in reflex as my forefinger pushes past, bulldozing the landscape around the tiny creature. I lift the jar, hold it close. The spider unfurls, crawls in. It is safe.

19 July
The last collection of yellow rattle seed lies on the sheet ready to be bagged up.

My kitchen scales tell me there are 619 grams.

20 July
I lift the scythe from its resting place in the shed, carry it to the field and soon the chiming of the whetstone rings out through the thin summer air.

22 July
There is something so delightful in steadily cutting this patch of earth that not so long ago was littered with blackthorn, that now is becoming a meadow, bordered by the dead hedge of that same blackthorn from which newly homed wrens dart as I slowly pass by.

It is hard going in places. The larger stubs of the bow-sawn blackthorn bushes catch on the blade alarmingly.

RENATURING

24 July
A glorious quote from Gerard Manley Hopkins from this day in 1871:

> Robert says the first grass from the scythe is the *swathe*, then comes the *strow* (tedding), then *rowing*, then the footcocks, then *breaking*, then the *hubrows*, which are gathered into *hubs*, then sometimes another break and *turning*, then *rickles*, the biggest of all the cocks, which are run together into *placks*, the shapeless heap from which the hay is carted.[18]

I love that insight into the lost vocabulary of scything.

27 July
This first patch of meadow is scythed. I have raked the cuttings into a hayrick of sorts – a thick stick stuck into the ground then heaps of cut grass placed upon the pole. They settle upon each other such that the rick is now some six feet tall and resembles a strange harvest yeti resting in the field.

1 August
In this year of extreme drought, there is welcome cloud this morning – a touch of dew.

In the field, dark dots in the oaks mark the emergence of young acorns. Last year, there were none. The oaks operate in a different realm to us humans.

I halt on the path as a sparrow-like bird flies before me and rests on a stem of ragwort. Brown-streaked, yellow head – it is the first yellowhammer I have ever seen in the field. I am ecstatic.

3 August
On my journeys beyond the boundaries of the field, I have met many fine people who are passionately engaged with their own

efforts to help bring back nature. It is always a delight to learn of some of the many individual humans quietly doing their bit. But I also know that so much of the work to restore nature is best done not by humans, but by other creatures that live in the landscape. By bringing back key species that have been removed from particular places and spaces – often by the violent actions of humans, it must be added – those lands can be renatured far more effectively. The beaver is a perfect example of one such key species.

Once native to Britain, beavers are amazing for ecological restoration of wetland projects. Put into a suitable environment, they get on with what they are best at – building dams, creating a series of pools and so essentially managing water in a stream-fed landscape in a better way than allowing the water to simply pour away. As such, beavers can be extremely useful at helping with flood prevention. It also means they very quickly construct wet, marshy worlds that are rich and diverse biospheres of pond life.

A while back I met Darren Tansley of the Essex Wildlife Trust. He is one of those impressive people who seem to have lived many lives. My friend Selfie had delightedly told me of Darren's previous existence as a rock musician, of his time in the somewhat legendary punk band Blyth Power.

'Always wore a battered old black top hat,' Selfie said.

Darren still has his impressive mop of black hair, now tethered in a ponytail. When we first met, we happily chatted on the wonders of aged oaks for a while, then Darren told me of his current work as river catchment coordinator. He was overseeing the controlled introduction of beavers into a site in North Essex. If I wanted, I was welcome to come and visit.

So today I get to meet up with Darren once more – this time on the green of the idyllic English village of Finchingfield on a glorious summer's day. It feels a good time to step out of the field and see some other projects aimed at restoring nature in my local area. Wear wellies, Darren had suggested. I get out of the car and change footwear. With Darren is his colleague Natalie Singleton, a river catchment officer who is carrying a wooden box with all their kit for eDNA testing. They are here to look for signs of water vole activity.

We wander away from the village centre, off down a lane that soon leads to fields. By a stream, we stop on the bridge.

'You've got Finchingfield Brook that runs down over there,' says Darren, pointing over his shoulder. 'And then this one that runs directly out of the beaver compound. What we've noticed is that that brook dries up every year. This one used to too. But since the beavers have moved in, this one now stays wet all the year round.'

'Ah,' I say, already impressed at just how effective the beavers have been at managing the water here.

'They're basically releasing the water they've captured in the winter, slowly, all the way through the summer. So when you're in drought, they're actually providing water, and when you're in flood, they're preventing flooding. They're a major water conservation tool.'

The local landowner who has enabled this beaver scheme is Archie Ruggles-Brise.

'His family have been here for three centuries. He's one of these people who has a lot of land but is absolutely behind any conservation initiative,' explains Darren.

'Nationally, he's quite an advocate for landowners getting involved in these kinds of projects. At the moment, he's really involved with forestry-farming. Plots of trees amongst the crops.'

'Ah, agro-forestry,' I say.

It really isn't long ago I was in that barley field in Suffolk learning of such ways.

We head on, up towards the entrance to the compound on the side of the hill. There is six-foot-high mesh fencing all around, with an electric wire across the top.

'This is unnecessary really,' says Darren.'

'Perhaps more to reassure some humans?' I suggest.

Darren smiles in agreement.

'Archie said he'd put in something that was all bells and whistles, just to prove the intent.'

He points out one of the design features – the wooden posts are on the outside of the fence rather than the inside. I laugh. Beaver-proof.

'That's the sort of thing that can easily happen, isn't it?' says Darren with a wide smile. 'You get a contractor who bangs in the posts . . .'

'And no one thinks about it . . .'

'Until the beavers have taken them down!'

In truth, the beavers never come up here to the edge of their enclosure. They stay down around the stream.

I ask if the locals have reacted well to the presence of beavers in their landscape.

'They love it,' says Darren.

The main thing for the village is the reduction in flooding.

Darren unlocks the padlock, then teases the contact point off the electric wire. He opens the gate and we step into the beaverlands. Their enclosure covers an area of four hectares or around ten acres.

A few yards down-slope, we halt before a swampy, sylvan scene where a wooden hut looks out on a wide pond. There is water and greenery all about.

'That's one of the hides,' explains Natalie. 'The beavers have dammed around it, so it's in the water now.'

I laugh.

There are a series of metal ladders placed on the ground to allow Darren, Natalie and other human visitors to move around the site.

'We started with wooden boards,' says Darren, smiling.

It takes me a moment to see the issue – the beavers simply turned them into material for their dams.

I laugh again.

A pair of beavers was released here in March 2019. Since then they have transformed the landscape along what was once a stream and now is a series of ponds, verdant reed-banks all clearly abundant with plant and insect life.

'The first project was down at Ham Fen in Kent in 2000,' explains Darren. 'That enclosed quite a large area and was very experimental at the time as no one had tried it before.'

He smiles.

'Some of those beavers did manage to get out during a flood. They went straight out over the top of the fence.'

I laugh quietly.

'They now live wild in Canterbury. Very few people know about them. They don't dam the river as it's too deep there. They just burrow in the bank.'

It's a lovely story. The notion of beavers settling in urban spaces on the quiet is a delight. But they are now being introduced more widely in our cities. As part of the Rewilding London initiative, two beavers were brought into an enclosed area at Forty Hill Farm, Enfield in March 2022, bringing them back to the capital after 400 years away.

We walk upstream towards the beavers' lodge on the main pond they have formed.

'All of the water you see would not have been here if the beavers weren't here,' explains Darren. 'It was just a single, straight brook running down. They have dammed it and backed the water up.'

Before us is a wide, open pond thirty or forty feet across. Then there are other smaller ponds that have formed as the beavers cut down small trees and make dams. The result is a huge volume of water held on the land and a steady flow of water seeping through all year round.

'That's their lodge,' says Natalie, pointing to a dome of logs and sticks six feet high on the far edge of the pond. 'The entrance is beneath, under the waterline. They dive down in and out.'

I can picture them. Cute faces above the water, wide flat tails propelling them. Not that I'm likely to see them.

'They're crepuscular,' she explains, 'so they only really emerge at dusk and dawn. For much of the rest of the day, they're tucked down, hidden away in their lodge.'

'Nice bit of beaver sculpture there,' says Darren, pointing to a small trunk of wood sticking out of the pond. The bark has been entirely removed, the beaver tooth-marks obvious when you look closely.

'Slightly abstract,' adds Natalie.

The lodge is probably the biggest one in Britain. They had previously built one elsewhere and then abandoned it and switched to this pond.

'This is better as there's a spring line that runs into this pool. Down there it was getting drier. This is the first pond they created. So they backed the water up in stages,' explains Darren.

'It's like having a starter home and then having kids and needing more space,' says Natalie.

That first lodge was built for them.

'They didn't appreciate that at all, did they?' she adds with a giggle.

'No,' agrees Darren. 'We built a little pool so they had some deep water, dammed the stream there so it would fill up and put a hay bale lodge in.'

The beavers didn't fancy it. It's a lovely comic image, imagining the beavers' reaction.

'We're not staying here!' I say.

'That's not fit for purpose, mate,' says Darren.

So, as Darren's explained, the beavers just swam downstream and started to make their own lodge.

'They located where the water was coming in from the spring and then diverted it into the stream and dammed the stream.'

It's so clever.

'They're actively thinking about movements of water.'

'Yeah,' agrees Darren. 'And how to move it around the site.'

'They're hydrologists, engineers and builders,' says Natalie.

The beavers should live for seven or eight years. There are few threats to them here in this enclosure. Wild beavers would have a reduced lifespan. These beavers have already had two sets of offspring.

Above us there is the distinctive fluttering of poplar leaves. Many of the trees will provide the main body of this beaver world.

'You do see little groves where there are lots of pencil-topped tree stumps,' says Darren.

Over the first couple of years, the beavers made substantial changes to the site.

'It's more established now,' says Natalie. 'But they're always making adjustments. They want to capture all the water they can.'

We've arrived at one of their dams.

'This is one they slung across quickly in the winter,' says Darren. 'They start off just by heaping up mud. Then they add these smaller twigs . . .'

He points to the mesh of sticks within the mud.

'Then gradually these larger ones go in,' he continues, 'and then you end with great big logs that they lump in to create a really strong dam across.'

'Bit like wattle and daub, isn't it?' says Natalie.

It really is – and a kind of messy but extremely effective way of capturing water.

'Normally, they just put a very quick dam up to capture rainfall,' says Darren. 'As soon as the dam starts to fail, they'll go in and patch it up. If they can, they drive the dam sideways to capture more water. If they need to, they increase the height of the existing dam.'

Yet each dam is actually slightly leaky, such that there is always a steady flow of water within the system, a gradual stream downhill through the mesh of sticks and mud that forms each dam, from one pool to the next, one great pond to another. That is the genius of this waterworks, what keeps it healthy and alive.

The fluid nature of the landscape is clear, as is the industry of the beavers.

Darren and Natalie haven't forgotten they are officially here to check on water voles.

We walk on, looking for the best place to test for their presence.

'Or mink,' he adds.

'That's bad, yeah?' I ask.

'Yeah.'

'Would beavers have an effect on them?'

The truth is, no one really knows.

'Beavers were already extinct here when mink came to Britain in the 1930s,' says Darren, 'so there's no precedent as to how they're going to interact. But I doubt mink will be much of a problem. Beavers are thirty times the weight of a mink.'

We reach the upper reach of the stream. The channel is completely dry. It's a vivid reminder of the effect the beavers have had.

'If we were relying on rainfall, this is what we would have,' says Darren. 'All of that water down there is because of the beavers' dams.'

'Wow,' I say. 'That's quite an impact then.'

It really is. As Darren and Natalie chat together about where best to look for signs of water voles, I have a moment to muse on the impact that beavers could have across Britain – not only alleviating flooding but in greening up watercourses, creating vibrant areas of reed beds, and strings of ponds and pools along streams and rivers.

'It's the retention of water within the system,' I say, pretty much to myself. 'You've got so much more life. I mean, it's dry as a bone in the fields at the moment.'

It's true. There's a heatwave, the worst drought for fifty years, and yet I'm standing in a world of water, one full of wildflowers, insects, birds . . . and hopefully water voles, too.

'That tree there has been felled by beavers,' calls Darren from where he has wandered to.

I head over.

There's a poplar tree that has been brought down by being gnawed at a few inches from the ground. The trunk must be eight inches across.

'It's a decent size,' I say.

'Yeah,' says Darren. 'They've brought down some pretty big willows, too. What they tend to do is knock a tree down, then take the main branches off and use them, leaving the main trunk sitting there as it's too big for them to carry.'

'Ah,' I say.

'The other thing is that they eat the bark anyway. It's food for them,' he explains. 'You see it all along. You find stripped branches with no bark because they've eaten it all.'

I laugh. There's something so funny about the image of a beaver sitting munching its way through a poplar tree. Funnier than my rabbit-girdled apple tree in the field, anyway.

'That's all they eat, isn't it?'

'No, they eat other vegetation,' says Natalie.

Beavers are vegetarian. So they can't be blamed for depleting fish stocks, like otters are.

As we walk on, Darren explains how otters have been in the fossil record of these islands for millions of years, how it's only in the very recent past that they have been absent from our rivers.

'They're not an introduced species,' adds Natalie.

It's such a good point. Just like the beavers. Just like the white eagles and, indeed, just like wolves. But unlike grey squirrels, or mink, or crayfish. Those have all been brought into Britain. They never existed in these isles until some dumb human decided they would play God and bring them here.

I follow Darren and Natalie around the edge of one of the large pools formed by the beavers, stepping carefully around the wet ground, marshy in places, with its green clumps of sedges and reeds. We reach a dam. Above us is another pool, though holding less water.

'There's probably more than six feet of water below here,' says Darren.

He's stood on the other side of the dam, looking down the stream.

'It's like a series of steps,' he says. 'Each one gradually drains down into the next one, then into the next. So they've got their store of water up here, to keep what they need flowing down there.'

Darren points back down to where the beavers are stolidly asleep in their lodge on their home pool.

'The biodiversity increase must be huge,' I say.

'You've got freshwater shrimp. You've got kingfishers come through. You've got water shrews,' says Darren.

There are currently loads of tiny common frogs all around us.

'You'd think the grass snake population would have benefited.'

I'm standing by a severed willow trunk from which three-foot-long stems of some strange plant have grown. Thanks to my friend Frances Mount, I know what it is. There's some in my garden.

'That's figwort!' I say rather overenthusiastically.

'You get this cascade effect of biodiversity,' continues Darren. 'Having that draw-down of water is so healthy. A single, sterile body of water is no good to anyone. There will be loads of stuff in there.'

He indicates the muddy base of the drier pool above us.

'And then, once winter comes, it will fill up,' I add.

'Yeah,' agrees Darren. 'And it'll be ready to fill because the dams are in place. Doesn't even have to be winter, it could be a summer storm.'

This is that other wonder of having beavers on your watercourse.

'Imagine the people in Finchingfield right now. If a summer storm swept in before the beavers came here, all the stream would suddenly empty into the duck pond at the bottom of the village. That fills, then the road goes under water, and then the houses flood . . .'

Now a rainstorm is nothing to fear.

'It will take ages for these ponds to fill up. A summer storm will make no impact because the potential to store water is so great. The stream will just continue to flow as it does now.'

I tell of my chat with Hazel at Wild Card, who is a volunteer flood warden at Hebden Bridge. Darren has friends who live there so knows the issues.

'The flooding is massively impacted by having a grouse moor above. If you were to change that to a beaver enclosure, I'm sure you could sort flooding much more naturally,' I say. 'But then you step into human ownership issues.'

'The landowning classes do not want to see that,' adds Darren.

'Nope. They want to shoot little black birds.'

It's a distinct sour note, but the ridiculous vibrancy of the figwort soon brings back my joyous mood.

'This is amazing,' I say. 'I've got one of these plants that's about this big.'

I hold my hand a foot from the muddy surface of the pool.

'And I've been looking after it all year.'

We step back towards the safer ground of the stream's edge. I can't help but wonder why beavers really aren't being seen as a better way of easing flooding than way more expensive and ineffective approaches like putting down physical blocks or dredging the rivers, which can actually exacerbate the problem.

'The Environment Agency is a partner in this project,' Natalie explains. 'Primarily, this is a flood-management scheme, as well as looking at biodiversity impacts.'

The Agency is monitoring flow in this stream, compared to the dry sister stream we saw earlier which has man-made dams.

Natalie smiles. 'The water levels are so high here thanks to the beavers, they've had to move their monitoring stations,' she says.

We laugh.

For Natalie, her clear delight is in seeing the transformation of the habitat and the positive impacts that has on the ecosystem.

'The oxygen flow is just so much more stable in the beaver wetland than in the man-made dammed stream,' she explains.

This is a pilot project – the first in Essex. There are others in the planning – spatial modelling helps with seeing what landscapes are suitable. But the basics are a stream and some trees.

'They're not going to dam main rivers,' explains Natalie. 'You're not going to see the Humber with a big beaver dam across it!'

'But they can be part of the solution to flooding,' I say.

With such clear evidence from a project like this, surely it's time to roll out more extensive beaver-based flood-prevention plans across the country wherever they are possible.

'It's early days, but there are certainly good signs,' says Natalie with a certain modesty. 'I think Archie wanted this as a blueprint for other landowners, to show them it can be economical, that it can have a benefit.'

Archie Ruggles-Brise – the enlightened figure behind the project – has made this happen by handing his lands over to the beavers. He is one of those key people who are enabling us to see another way of being, initiating dramatic changes to the environment even in these more populated parts of south-east Britain. By so successfully introducing beavers into the village of

RENATURING

Finchingfield, he's offering a new vision to how streams and rivers can be managed – both to benefit biodiversity and stop the threat of flooding.

'We're getting to the stage where people are ready to move beyond the fence,' says Darren.

I know what he means.

'At the moment, we're not quite there in Essex.'

Setting the beavers free of their enclosures here may be a little while away. There are parts of the country better suited for beavers, like the West Country, with its wooded valleys. But that's something I can dream about when I'm back in the field: a world where beavers are back all across Britain.

On Reintroducing Species

One of the most obvious ways to restore complex ecosystems to wilder, healthier ways of being is the reintroduction of certain central species that have become extinct. There are various species that we know have a vital role to play. Wolves, lynx, beaver, elk, bison, wild horse and wild boar are all what are known as keystone species. Reintroducing them into large-scale rewilding projects has a positive domino effect in bringing back a strong, varied and species-rich ecology.

Take the beaver, for example. Since being hunted to extinction in Britain some 400 years ago, they are now being welcomed back for the work they do in acting as wetland ecosystem engineers. Across western Europe and North America, beavers have been brilliant representatives for successful rewilding. Industrious and cute, their efforts at tree-felling, dam-building and generally transforming streams and rivers into thriving, dynamic systems are now widely recognised. Other keystone species might be less endearing but can also have great value for nature restoration projects. Large herbivores like bison and Konik ponies can have a huge impact on open rough grasslands, ripping out the vegetation while scarifying and carving up the ground with their hooves to form new worlds for seeds and insects to step into.

Certain species, such as wolves, are rather more controversial. Yet while some humans may fear their reintroduction, the effect of returning wolves to act as top predators, apex carnivores, cannot be ignored. Yellowstone National Park is a perfect example of how bringing back wolves has restored the ecological health of the region. Deer and elk no longer overgraze the land but are now more wary and move on more regularly, allowing plants, trees and flowers to return, meadows to rise. Willows are growing taller. River systems are recovering. More elk kills also means more for the scavengers. The cascade effects are being positively felt throughout the trophic system.

Good news stories are important. They give us hope. Ecological recovery can be led by reintroducing iconic species. Human actions may have erased them from their natural landscapes, but the act to return them can be both hugely redemptive to us and valuable to the ecosystem they vanished from. In Britain, the return of white-tailed eagles has been a wonderful success story. From their first sites of recovery on the west coast of Scotland, they now nest as far south as the Isle of Wight. Sometimes, the closest possible surrogate reintroduction can work, too. In Mauritius, the introduction of a species of non-native giant tortoise has acted successfully as a substitute for an extinct native version.[19] The local domed Mauritius giant tortoise (*Cylindraspis triserrata*) was slaughtered in great numbers by Dutch colonists, becoming extinct on the island by around 1700. Yet in recent years some 200 young large and giant tortoises (non-native megaherbivores, as they're known) have been brought to Mauritius and have proved excellent at munching away and managing the swathes of non-native plants to have invaded the island.[20]

7 August

It is a young writer called Jack Stimpson who puts me on to them. We are chatting about his plans for writing a book about trees and he says he is due to meet up with Youngwilders. It is an intriguing name. I don't know of them. He tells me they are all about connecting up young people who want to volunteer their help to bring back nature with those who own land and want to do some form of ecological restoration project.

It sounds such a neat idea.

So today, I step back from the field into a meeting with Jack

Durant – director and co-founder of Youngwilders – and Noah Bennett, their chief technical officer.

Jack is cutting an exotic figure in his Zoom window with bold moustache and mullet, smiling wide and beaming in from sunny Wimbledon.

'True sustainability is always going to have to be cross-generational. We can't just leave it to aged crumblies and let it all fade away. If you want that bigger picture of "earth thriving in perpetuity" stuff then you've really got to get young people hyped off their nuts for it.'

I can't help but grin back at the image of Jack before me, smiling in the sunshine sixty miles away. He and Noah embody the hopeful future of nature restoration. They are twenty-something and they care. Really care. It's there in Jack's smile – the passion, the enthusiasm to make a difference, to forge change.

It really is lovely to meet these people. After trying to get my head around the rights and wrongs of large landowners rewilding and how some financiers are clearly muscling in on rewilding as a neat way to make money, there is an honesty and conviction to Youngwilders that is a delight.

'Our approach is that we have to get cracking with this,' explains Noah. 'We're working on a small scale. We don't have any land. We don't have any money. But we need to get going with wilding the land. If we don't perfectly replicate what the ecosystem was 3,000 years ago, that's fine. We're still enhancing nature, building connectivity.'

'Your website is great,' I say. 'You use the word "wilding" rather than "rewilding". Why is that?'

'Wilding seems a less prissy word, perhaps,' says Jack.

We chat about how the term 'rewilding' should really only be applied to grand-scale projects where herds of large herbivorous animals roam free through a landscape shared with apex predators.

Yet such a definition excludes the vast majority of people from being engaged with ecological restoration.

'Let's get everyone involved,' I say. 'Otherwise you can't live in a city and be part of this.'

There's one more thing that I really want to ask about. On their website they use the term 'DIY-ilding'.

'It's a nice term,' I say. 'It's playful but also says, "We can all do things". Even if you only have a window box. You state how you've positioned yourselves as opposed to "the profit-prioritised rewilding projects" . . .'

'Yeah,' says Jack. 'That commitment will probably be our undoing.'

He makes a wry smile.

'Basically, I'd been speaking to a lot of crumbly conservationist kings and queens who have been in the game a long time. And the one thing they've been really clear to communicate is that this is a really unique time in UK conservation. There's massive public goodwill behind the conservation movement. It's very positive rather than defence-orientated. And money is being put where mouths are. That's totally unprecedented for conservation.'

Jack has studied conservation management. Like all the founders of Youngwilders, he knows his stuff.

'If we'd done this ten years ago, we'd be chatting to a few small landowners. Now, we've had conversations with fancy-pants banks. Kind of greenwashing-type stuff. Also land grab. Buying loads of land for the sole purpose of engineering it to maximise government payments, carbon credits, that kind of thing.'

It doesn't sound great.

Jack continues.

'Obviously, there's a lot of things that are worrying about that. It does mean there will be more rewilding, but it does really depersonalise what is a beautiful human movement. I mean, some think, "We've got this 10,000 acre site. We're going to get twenty professional tree planters to storm in and out." They're going to maximise trees per square metre but biodiversity concerns will be secondary.'

Youngwilders are about giving everyone the chance to get involved.

'Seeing people who love nature being part of it. I hate the idea of enclosing the process, as is happening at the moment. Giving people a chance to self-initiate and get stuck into their own desire is a much more exciting and society-fulfilling vision of restoring nature. Better than this technical, over-professionalised version that will happen in the next five, ten years.'

It's a serious point. Jack makes it so well.

'I really like that,' I say, then lighten the moment. 'What did you call those conservationists? The crumbling kings and queens?'

He laughs.

'I mean that in a really affectionate way,' he says. 'God knows what the UK would have looked like without their work. It was just fascinating to hear from them about the challenges they faced. In a sense they're bigger for us because of the ecological damage since the 1970s, though there are now these money levers which we could pull if we wanted. Now we have these huge institutional forces pushing us around, whereas before, perhaps, we would have had more agency. Today, it's a very different landscape.'

It's so fascinating to hear this.

'There's a market that is being set up by government policy,' says Noah with a calm clarity, 'which means there is money to be made. People are investing.'

He pauses.

'It is a good thing. There will be more biodiversity in the UK as a result of "biodiversity net gain". But where we come from, we want to see people involved, we want to see the community having a sense of ownership over these rewilded spaces.'

Jack speaks up.

'Yeah, exactly. We have a vision of these smaller projects being run by local young people. Then they will leave more connected to nature but also upskilled to have a full, long, satisfying career working in nature recovery.'

'That's so important,' I say. 'Community-led nature restoration of local landscapes.'

I'm intrigued to hear of other people or projects that Youngwilders work with who they're especially impressed by.

'Hannah Needham,' Jack says. 'Heal. They've been our long-term collaborators. Hannah is super-inspiring. They're a charity. So they dodge that profit-driven concern.'

'Heal,' I say, repeating the name as I have a quick search online.

'They're also very ambitious. They're outward-focused, trying to get as many people as possible involved. They have this three-by-three thing, which is pretty cool. So if you donate you get a what3words metre square.'

I've just found the website. I click the About Heal tab and read:

'Heal is raising money to buy land in England and rewild it. We're giving land back to nature, forever.'

'It's very people- and public-focused but still large-scale,' says Jack. 'I think that's awesome.'

Certainly is. Even a glimpse at their website shows their endeavour.

'The Langholm Moor Initiative,' says Jack.

He's moved on to another.

'We did a rewilding tour when we were starting out,' he says. 'We went to see them a couple of years back. They're in southern Scotland – a community buy-out of this landowner.'

This project I have heard of.

'We met the team. They were all so awesome. It's just so wholesome – top to bottom.'

Noah speaks up.

'One that comes to mind is Thames21.'

They're new to me.

'They're doing river restoration on a range of scales – all along the Thames Valley from the River Thames to the various tributaries. They do great volunteering days.'

Even though I know I'm never going to speak to all these fine people, it really is heartwarming to hear of their projects. There's a feeling of being part of a wider community of other souls, each engaged in their own ways with renaturing the world. My energies are best spent restoring a wilder ecology to a small field in Essex rather than dashing about seeing other enterprises, I tell myself. I'm not a journalist. I need to remind myself of that.

'Water restoration is something that does get overlooked,' Jack says, bringing me back to the immediate. 'And it's possibly the most devastated part of the UK ecosystem. I just remembered another group – they're called something like the Isle of Arran Community Trust.'[21]

A bell rings.

'Ah,' I say. 'I know. They were cleaning up Lamlash Bay.'

'That's them. They were super-nice. Total pioneers.'

'Yeah,' I add. 'They call it sea-wilding.'

'Ultra pioneering,' says Jack.

For a moment, I think back to a visit to the glorious Isle of Arran off the west coast of Scotland. Then we start to talk about the ways in which Youngwilders and other smaller nature restoration organisations can get publicity for what they're doing.

'An ambassador?' I say. 'Leonardo DiCaprio?'

Jack laughs.

'Yeah, look him up,' he says.

'You could hook up with him?'

We all laugh.

'It must be a balance,' I say. 'It can't be easy to oversee your growth as a project.'

I suggest that the singer Ellie Goulding might be a good person for them to have as a figurehead and show a newspaper article I've squirrelled away with a quote from her saying, 'There's a YOU-shaped hole in ecological activism!'[22]

'Oh, no way!' cries Jack. 'I heard she was into it. That's awesome.'

He tells of a friend who does foraging walks and how Kate Moss went on one and was saying she's getting into rewilding.

'Our friend Issy gave her the Youngwilders contact details and then told us,' he says through loud laughter. 'We were waiting by the phone.'

'You never had the call?'

'The call never came. Maybe it's better to avoid associating with huge names,' says Jack. 'Though I can't quite articulate why!'

'Down the line,' I suggest. 'Once you're more established.'

'Yeah,' says Noah. 'I think we're still holding onto our radicalism for as long as we can. Being open to making mistakes. Learning some lessons while we're still a small organisation.'

'The stakes are lower,' adds Jack. 'We've been navigating this for only two years or so. I guess the map will form as we stay on the scene.'

We've soon wound back to the contrast between idealistic community-led projects and the presence of business-minded companies stepping into the field.

'When we do talk to them the single 2D-line metric they're operating on is "will this lead to more money?"' Jack says. 'And inevitably there's so much more happening that is overlooked – and the stuff that gets overlooked is so vital for saving the earth from ecological decline.'

He speaks so well on this difficult issue.

'I just don't have the faith in the market that some other people might,' he adds.

That's exactly the problem. His doubts are mine. You need people who are involved because they want to improve the environment – not just to make money.

We start to say our goodbyes. I step away from the computer, change into field gear and head out the back door seeking my scythe.

8 August
In warm sunshine, I simply stand and stare into the face of a corncockle flower, savouring the beauty of those purple petals. It is one of the last of the season. Others have curled tight already. I am due to scythe this patch of meadow but am held for now by this single splash of mauve. Nothing of such wonder should be felled before its time. So I stare instead of scything.

9 August
Hidden here and there among the thick grasses of the field are more sprays of common centaury, with their tiny pink flowers.

Centaury

11 August
Too hot to scythe.

RENATURING

15 August
Night is the time to venture into the field. In the moonlight, the ragwort seems to have run riot. I weave along the deer path savouring the stillness, the cool touch of the air. I stop and gaze upwards. Tomorrow I shall head away north for ten days – to camp out in the wilds of Galloway. It will be hard to step away from the field for so long.

1 September
My friend Chris Gibson, the ecologist, came up with the name 'Rewilding the Mind'. It is to be a symposium, a conference run by Beth Chatto Gardens, organised by Julia Boulton – Beth Chatto's granddaughter – who I've met a few times now. She kindly invites me to come along.

When exploring the notion of 'Rewilding the Mind', Julia states 'our intention is opening our own minds and others' minds to giving nature a stake, a voice in everything that we do; championing nature, supporting nature, underpinning nature'.

So now I sit in a vast amphitheatre. There is a buzzing murmur of human voices. The hall gradually fills. Others sit in the empty red seats about me. We each have a paper badge with our names written upon them that makes me feel we are like children gathered there excited, anticipating the events to come. A hush falls.

Arit Anderson – garden designer, writer and presenter – steps forward before us, welcomes us, tells of her own excitement at being here, and her smile, her voice, let us know this is true.

Then the speakers come.

Dan Pearson – landscape designer, horticulturist – tells how we have to 'relinquish control' over the natural world. I nod in agreement and scribble a note. Alastair Driver – director of Rewilding Britain – calls himself a 'nerdy naturalist' and speaks of influencing government funding policy. Then Adam Hunt, who co-designed the recent award-winning 'beaver' garden at the Chelsea Flower Show, appears and speaks of the emotional reaction of audiences to that garden, of his striving to forge feelings of 'awe' in the natural spaces he works upon.

At the break I step out into warm sunshine, awash with words. I feel a sense of delight and wonder that so many others, in their own ways, are seeking to restore a wilder, healthier ecology in the world.

5 September

I watch two jackdaws stood upon a telegraph pole, surveying the lands around them. Even sat here at the kitchen table, I am a part of nature. I think that is the vital starting point for all thoughts of renaturing.

A young male pheasant appears on the patio only a few feet away. He is a teenager, ungainly and unkempt, freshly raised from one of a crop that had been placed in the woods around, to be shot once he reaches a more mature, plumper, easier-target size. His red cheeks aren't the full crimson yet. His many-coloured coat is rather ragged, still growing. He looks nervously around, stepping over the sandstone slabs, and I think how these really aren't his natural lands. Pheasants are not indigenous to these parts. His species was introduced to England – brought here centuries ago, raised in vast numbers from the nineteenth century on English estates as a game bird.

I watch as he pecks at the ground, then catches a creature as it flies from the grass, a moth perhaps, and think of how these introduced pheasants have a hugely damaging effect on insect populations, not to mention lizards and snakes. Surely renaturing is also about questioning the value – the acceptance of pheasant-shooting, of grouse-shooting – of changing entire landscapes purely for the pleasures of a few gun-toting men?

RENATURING

7 September

Even as I worked happily away in the field, clearing ground of bramble, forging space for wildflowers, there remained a niggling doubt. It was to do with a worry that much of the good of large-scale rewilding was being quietly eroded. I had stepped beyond my little field and seen something of the ways in which some large landowners in Britain seemed to be profit-prioritising nature. Rather than genuinely seeking to improve biodiversity through rewilding, regenerative farming and the like, the drive for green capital seemed to be key to some projects.

I needed to hear the stories of landowners who are about restoring nature not merely to make money but because they care heart and soul.

Fortunately, I had been invited to come and see one of the most benevolent and largest projects of nature recovery in England. The Somerleyton Estate encompasses some 5,000 acres spanning the Suffolk–Norfolk border. Hugh Somerleyton is the current incumbent of these lands. He kindly suggested I meet him 'at the hall'.

So, one Wednesday in September, after dropping Molly at the school gates in Halstead, I drive out of Essex, over the River Stour at Sudbury and east across Suffolk. As I settle into the journey, it dawns on me that I am actually heading to meet a Lord and Lady at their ancestral home.

Unfortunately, when I eventually reach the village of Somerleyton, my satnav sends me to a gated entranceway that is locked. I drive back around the walled estate – past a rectory, a series of ancient oak trees – following signs to 'parking for coaches', and eventually into the estate proper. I get out of the car and look about. There is a wooden doorway in the high brick wall. It is locked. The only person around is a gardener working away within a neatly hedged enclosure in the distance. I am now late. Over a cattle grid, the road continues – marked by a 'private' sign. I jump back in the car and follow the road, curving round past another ancient oak, then opening out to the sight of Somerleyton Hall before me. There is a collection of children's bicycles and 4x4s scattered around the front of the hall. I park up. There is no sign of anybody. In the rather imposing front porch lie a pile of wellington boots. The grand wooden door opens onto a lobby. I step in. Two polar bears stand before me at full height – stuffed, white and frozen, and staring black-eyed down at me.

'Hello,' I call out. 'Is anyone there?'

I am feeling rather Alice-like again.

These are not the only bears in the room. There are smaller, browner versions – also stuffed – and on the far wall, the skin of what might have once been a white tiger. I step forward to the next door whose brass knob turns in my hand, letting me into the entrance hall. A wooden staircase runs away to my right. There are children's toys on the floor. It is starting to feel rather invasive, entering into this place which is not merely an English stately home but also clearly a family home.

There's a noise to my left. I open another door, call again and am greeted by the kindly, smiling figure of Lara Somerleyton, the lady of the house.

'Hello,' she says.

I smile back.

'Hello.'

It is a relief to find somebody. I am apologising to Lara for my lateness as Hugh appears.

'Don't worry at all,' he says.

He is looking for his boots. He is also after his wife's keys.

'We'll take your jeep, if that's OK?'

Lara looks at me and rolls her eyes.

'Sure,' she says.

A gorgeous one-year-old lurcher unfurls from a dog basket and trots over to greet me.

'And Paris,' he adds.

Hugh wants to show me as many of their efforts at nature restoration on the estate as possible. We jump in an aged 4x4 in the driveway. These 5,000 acres were handed over by Hugh's father some twenty years ago. Now, Somerleyton is becoming a model of regenerative farming.

'Wild edges, I like to call them,' Hugh says. 'It's in this scrub where everything lives.'

We have bumped our way along to a spot in the corner of a field. Paris and Achilles, a more sensible Patterdale terrier, lean their heads over from the back seat. Hugh explains how leaving a few yards fallow and allowing boundary hedges to thicken and flourish can have a huge impact on the biodiversity of farmed lands.

But we are also here to see something of the 1,000 acres of the estate which Hugh has turned to full-on rewilding. The story of his engagement with nature recovery is fascinating.

'Two years ago, Olly, Argus and I drove together to Scotland,' he explains. Hugh's talking about the other two founding figures of Wild East – Oliver Birkbeck and Argus Gathorne-Hardy – both also large landowners prepared to commit at least 20 per cent of their estates to wilderness.

'Ten hours from here. To visit a rewilding project – an inspiring guy called Paul Lister.'

'Ah, the wolves man,' I say.

'Yeah, the wolves man,' agrees Hugh.

Paul Lister is a conservationist and owner of Alladale Wilderness Reserve – 23,000 acres of the Scottish Highlands. He is best known for his desire to reintroduce wolves.

'And it was really the journey there and back that changed us. We're all privileged, educated and eco-anxious. Our big thing was, how do you make nature recovery alluring, dynamic and exciting – emulating some of those great rewilding projects like Yellowstone,

Carpathia but without bears, wolves or snow-capped mountains – in flat, rather bleak, East Anglia?'

The track we're now driving down seems to have vanished. Paris, the young lurcher, scrambles into the front, deciding my lap is a better place to sit. Achilles stays in the back. We bump along.

'Out of that thinking came the idea of democratising nature recovery,' he continues. 'It's hard to connect rewilding and returning big, iconic species with everyday life. It's not very sexy, what we're doing, but it is the tiny voices becoming one that is so exciting. We are richer by the sum of our parts. Backyard, schoolyard, churchyard, farmyard. "Thousands of micro-interventions in a macro-scale"; "The Facebook of nature recovery" – these are Argus's terms. The idea is that it becomes societal and leads to lasting change.'

Those are the essentials of Wild East's vision. Set up as a charity, the project is open to everyone prepared to pledge to renature 20 per cent of whatever wild space they have – be it 1,000 acres of an estate, a patch of suburban garden or a window box. That 20 per cent is a key figure – it's the minimum threshold for life to thrive in the landscape.

I tell Hugh of the chat I had with Youngwilders last week.

'They were saying, "Wild East? They're super awesome!"'

A shy smile creeps onto his face. I turn to the matter I really want to hear his thoughts on.

'Everyone knows we need the scale that large landowners can give,' I start. 'But I have heard the concern that the ideals of rewilding and nature restoration are going to get washed down the river as the money comes in.'

'That's an interesting point,' agrees Hugh. 'And entirely accurate.'

He tells me of how when they set up Wild East, they had the hope of acting as an interlocutor between 'thousands of micro-interventions' – the individuals who had signed up to renature 20 per cent of their garden, window box, etc. – and large industries who were looking for carbon-offsetting projects.

'The problem is that these people are genuinely doing their bit for nature recovery, yet because their landholding is not registered they don't qualify for a Natural England grant. They don't get noticed. The danger is that any finance goes over the top of these people, who are doing important work and aren't even getting paid,' adds Hugh.

'Say that window box is a hundredth of a hectare,' he continues. 'Its biodiversity and carbon sequestration are potentially very high compared to the same area in a field of wheat. One is high in food, one in terms of diversity.'

The comment is so insightful. It leads to a discussion on farming, food wastage and the ways in which farmers like him need to be given a slightly different message.

'Society is asking us to use the same zeal we currently have to grow food, to now grow nature and food.'

Nature and food.

I think of that Suffolk farmer at the FWAG evening earlier this summer standing in a field on the Helmingham Estate down the road, telling how sixty years before, his family was being asked to rip out hedges to maximise field sizes for crops.

Hugh points to the hedges, those wild edges that are such a vital starting point for this shift in farming.

'If we all gave ten metres, we would return 5 per cent – 70,000 hectares of Wild East – back to the wild.'

They are startling figures.

'If we're successful, that network could become the first lineal national nature reserve. Rather than the Dales or any of those other very important places which are isolated, this Wild East version would be a highway for all living things.'

Hugh's passion is so obvious. But he also speaks sense.

'Farmers will say, "I just want to protect my culture". I mean, I live for the land. I totally agree. I would do anything to preserve that. But let's not forget that when those coal miners, steelworkers, car manufacturers – and all those that were the backbone of British industry – were swept away from economic expediency, their cultures were also wiped out. It wasn't just pits that closed; the culture of those pit families was lost almost overnight. There were those whose father's father had been a miner. Just like that farmer will speak of how "my father's father was farming here".'

He steers us out of the field onto the main road.

'The only reason we farmers are here is because we've been ring-fenced by grant subsidies. Otherwise, that culture would also have gone. We're here at the gift of the taxpayer. Now, we need to farm for nature, to restore biodiversity which we have inadvertently destroyed. If farmers don't get that, then they don't deserve to be the recipients of grants.'

He laughs.

'That's my militant part. Most farmers are authentic and care a great deal about what they're doing, but we have sleepwalked as a society into this intensive farming vicious circle – cheaper and cheaper food, more and more of it wasted, worse and worse animal welfare, unhealthy environment and biodiversity, and unhealthy humans.'

It's a stark picture but one that is so true.

'We're killing a billion chickens. And if you think 30 per cent is wasted, that's 300 million chickens wasted. That's just insane. The alarm bells should be ringing. Is that what successful humanity looks like?'

He glances over as he drives.

'The big wake-up is the climate emergency.'

Paris shifts on my lap.

'Through the shock of climate catastrophe, the fundamental

question is: can we — the white, Christian capitalists that dominate the business world, the hard-farming world of North America and Europe particularly — learn to evolve, learn to take our foot off the accelerator and make space for nature? Can we actually accept that we didn't get everything right?'

He lifts a hand from the wheel.

'Those indigenous tribes which we are now beginning to relearn from — North American, Amazonian, Aboriginal and African — we're now finally seeing they knew about fire management or holistic land management — those peoples that we had exterminated, swept aside . . . can we learn to become enough like them to restore the ecosystem which is the world we live on?'

We bump along.

'It's a fundamental thing. Since the 1950s, we've lived through a time of plenty — fridge freezers, processed food, this wonderful and celebrated time of plenty — and I'm not criticising it. Most people have grown much richer. We can order whatever we want. In the words of Mark Carney in his Reith Lecture, "Do we want the Amazon that delivers everything, or do we want the other Amazon?" That weighing-up of who we are is really interesting. The point is, can we moderate our ways?'

I have to admit I'm a little stunned. Not at the truth of Hugh's words, more just at hearing them as we bump along here in a field in wildest Suffolk. There is a jolt within me of what I can only call hope. If this man beside me who owns such a sizeable chunk of prime farming land in East Anglia can see the truth, maybe there really is hope for the future. Maybe we're not all doomed.

'It's so liberating to hear you saying that,' I say. 'Because there is a worry that there are many large landowners that see rewilding merely as a new way to make money.'

'Absolutely,' agrees Hugh.

'But what you're saying is really different.'

He starts to speak further about the need for seeing complexity.

'A few years ago, rewilding was pilloried for being a rich man's game. You had Knepp and a few of the other kings of rewilding. That's not to knock the good they're doing – they could be mining with their money – but there's some way to go if we're to bring together the people-up, community-led projects along with the top-down ones.'

We turn off into a field tufted with ragwort. Paris lifts her head to the window.

'There they are!' Hugh calls.

I'm bemused. I follow the direction of his hand but can't see anything.

'Water buffaloes,' he says, and turns the jeep towards them.

The shapes of some large animals emerge in the far corner of the field. As we approach, Hugh steers the jeep round in a tight circle so that I'm closest to them.

'Wow,' I say. 'They're amazing.'

I lean out of the window. The buffaloes lift their heads somewhat nonchalantly – great black ears flick to fend off flies. They glance up and then return to grazing. The curve of those thick, solid antlers tells me they are no cows. With 1,000 acres to roam, they will help the rewilding of this corner of England – ripping and scratching at the ground, tearing up grasses, exposing the earth, opening these fields to other life. And these buffaloes are just the start. There are bison and beaver introductions also planned. There are already Exmoor ponies, Highland cattle and large black pigs doing their bit to engineer a natural landscape not seen here for thousands of years. Together, these animals will oversee the re-emergence of an uncultivated, biodiverse and dynamic environment – a real Wild East.

RENATURING

8 September

In the field, I tuck down among the green haze of the young oaks. There are so many oak galls this year – not the normal marble ball type but these ridged growths of oak knapper galls. Back inside the cottage, I learn there are some seventy different types of oak gall, all caused by various gall wasps laying their eggs in the tree. The strange swellings of the galls are a result of the oak reacting to the presence of the wasp larvae in spring. Apparently, the oak knapper gall wasp only arrived in Britain in the 1950s. In the field, they seem to be thriving.

10 September

In the wilds of the western edge of the field I find a Robin's pin-cushion on a dog rose. It looks like a tangled ball of hair but is actually another gall – one made by a rose gall wasp.

14 September

Two sounds reign in the field. Above, the distant mewing call of two buzzards, up among the clouds. Here, on the ground, the metallic chiming of the scythe blade – ringing out with each brush of the whetstone.

Neither I nor the scythe are used to this type of work. There is no easy swish of the blade slicing wildflower stems from the earth. Instead, this patch is thick with grass. I slide the scythe into green swathes, slash at the ground.

Now, rooks caw. Pigeons coo.

18 September

I step into the field. I shall scythe. I have a new patch of meadow to mow.

A small puff of powder lifts from the whetstone with each stroke on the blade, accompanied by a ringing chime. It is a beautiful autumn day. I sharpen the blade and turn to the meadow, grip the scythe with two hands and sweep the shining arc over the ground.

For some time, all I think of is that action – keeping the blade moving just above the earth, back and forth, back and forth. All I see is the grass, the dried casings of yellow rattle, of thistle.

Then I halt and look up from my task, called from above, from the pure sky, by two buzzards whose pale undersides circle and dance against the blue. I watch. They have drawn me away from my work and now I see the life around me. Wrens on the lichen-littered blackthorn. A dunnock darting over the ground. A robin fluttering around the young oak, the wild rose.

These serene moments are my rewards for renaturing the field.

20 September

I rub a pod of corncockle between my fingers and collect half a dozen of the round black seeds in my palm. They are so much larger than most other meadowflower seeds. They must have been fairly easy to 'clean' from the wheat fields once their poisonous nature was recognised. Yet in traditional medicine that same toxicity was what made them effective for curing warts.

Corncockle

RENATURING

21 September

On the drive up to see my son Joe in Nottingham, I once again listen to Alexis Nikole Nelson – aka @blackforager – talking on a podcast called *Extra Spicy*. She speaks in her glorious sing-song way about urban foraging, identifying and cooking wild foods like garlic mustard and dandelion flowers. She speaks about the racial history of foraging laws in the United States. It is a startling and spellbinding lesson. This feels like the best form of renaturing education. She tells how foraging is often seen as a 'white, upper-crust activity' and how that made her decide that she should be present, frame herself as @blackforager. Put simply, foraging has been white-washed while ethnobotany has become a buzzword.

'So many practices that have been passed down through generations of indigenous people might – oh my gosh – be worth some kind of merit,' she says with perfect irony. 'Who would have guessed?'

Alexis explains how so many of the laws disenfranchising indigenous people from foraging wild plants, from stepping into public spaces to gather food, were brought in around the time indigenous people were being put into reservations and Black people were being freed from the South. It is a shocking truth.[23]

She talks more about how people still practise what she calls 'concern policing', such as when she gets asked if it's legal to be gathering persimmons from a local park that would otherwise have fallen and rotted on the ground. The host, called Soleil Ho, says how within capitalism it can be very confusing for people to be getting something for free.

'It just baffles folks,' agrees Alexis. 'Folks will do mental gymnastics to come up with reasons why foraging should not be happening and people should not be partaking in it. There are good reasons.'

She pauses.

'A good reason would be if you're not particularly good at identifying plants! Maybe sit this one out if that's you.'

She's hilarious.

'Like it's OK that sometimes people just get free food by accident. It's fine.'

Somewhere near the junction for the M1, I turn to the next podcast I had lined up: *Rewildology*. I'd downloaded an episode called 'How Do We Rewild the World?' hosted by Brooke Mitchell-Norman, who promises to take a 'deep dive' into rewilding. Within a minute she is quoting from Steve Carver's 'Guiding Principles for Rewilding'. I seek the slow lane and listen.

'What's the difference between restoration and rewilding?' she asks.

She starts to talk about Y2Y, that conservation initiative to protect a huge corridor of landscape running some 2,100 miles from Yellowstone Park in the United States through to Canada's Yukon Territory and covering an area around 500,000 square miles in size. Proper rewilding.[24]

As I drive I try to imagine the scale of this project. It's hard. My journey to Joe is around 100 miles. Even if I were to keep driving all the way to Cape Wrath on the northern coast of mainland Scotland, that would only cover 600-odd miles. End to end, Y2Y is well over three times that distance. That is seriously large-scale rewilding.

I settle into the slow lane and listen on as Brooke skips across the world to the Highlands of Scotland to talk about restoring ancient Caledonian forests and reintroducing lynx and beavers. Then she turns to southern Africa and starts to explain the foundation of 'Peace Parks' in the 1990s. The largest of these is called KAZA, the Kavango-Zambezi Transfrontier Conservation Area, extending across five countries (Zimbabwe, Zambia, Botswana, Angola and Namibia) and covering 520,000 square miles, even larger than Y2Y.[25]

'Wow,' is all I manage to say.

There is something genuinely reassuring in knowing that there are rewilding projects of such scale successfully being undertaken in the world.

It seems an age since I spoke to Steve Carver and began to recognise something of the complexity of the matter – from the notion of spatially mapping countries and continents in order to define the best spaces to rewild, to the emergence of wildwashing by companies whose central desire is not so much to green the world, halt biodiversity loss and ease the climate emergency, but to make money. Of course, many would argue there is nothing wrong with having a desire to do both.

For some time, the realisation that there were individuals and corporations within the rewilding movement who were there only for financial gain had shocked and outraged me. I was naive, I know. How can you expect people within a system centred on capital to be doing things solely for the good of the planet? You can't. Yet that had seemed the ideal, the hope that so many nature restoration projects had been founded upon. It wasn't for the money, it was for the good of the planet.

Only yesterday, I had watched a recording of a fascinating webinar centred on guiding investors into ESG (environmental, social and governance) financing, with a specific focus on rewilding. Alastair Driver, the director of Rewilding Britain, spoke of how 'traditional nature conservation practices on their own are not enough to achieve significant wildlife recovery in Britain'. He spoke of how we have a climate emergency, how we need 'nature-based businesses'. Then Charlie Burrell, owner of the Knepp Castle Estate, took the floor.

'Our soils are pretty poor,' he explained. 'This is why we went out of agriculture.'

The results in terms of ecological restoration had been startling – from the appearance of white storks that decided to stay, nest and breed, to the clouds of purple emperor butterflies seen around the ancient oak trees. Yet the financial results were also impressive. Burrell outlined how in 2019–20 they had made £600,000 from eco-tourism and camping/glamping alone. These were encouraging figures for financiers considering investing in rewilding.

I had been left confused. Surely I wasn't naive enough to think the only way to successfully rewild the world was through goodwill? After all, we live within a capitalist framework. Private finance is required alongside public. Right?

Right.

Yet that was where a creeping sense of unease arose. Can we trust that capital investment into rewilding will do the right thing? Won't quick profit prove too tempting a reward? Just as the way in which carbon-offsetting by big business has led to all kinds of dodgy practices around reforesting projects – planting the wrong trees in the wrong places, not caring for them and merely paying lip-service to the entire practice. Won't similar problems arise within rewilding when the focus is on finance rather than fragile ecologies and a genuine care for nature? Who is going to ensure that doesn't happen?

I had skipped through the webinar to a moment when Charlie Burrell was asked about people's reactions to the notion of bringing back large predators such as wolves.

'The goal should always be to attempt to reintroduce more and more lost species that were part of our landscape, part of our ecosystem,' he said.

The words were reassuring. He was so right.

The more complex the ecosystem, the better, he explained. It was simply a matter of educating people on what such reintroduced species would do for the ecosystem. I thought of those deer shifting around uneasily at Yellowstone Park once wolves had returned. Charlie Burrell spoke of the work of David

Hetherington, who has so neatly explained the positive impacts of bringing lynx back for the ecology of these lands.[26]

His words had given me hope again. Perhaps even in my lifetime there would be lynx and wolves walking freely on the lands of this island chain, this archipelago where I lived. I would focus on my own, rather smaller-scale, efforts at bringing a more complex ecosystem to the field, and leave large-scale rewilding to the experts.

Anyway, for the moment, my podcast has finished and it is my junction. I turn off the M1 and head towards Nottingham to see Joe.

22 September
Gathering yellow rattle seed – inverting the grey, dry heads and watching the tiny circles of seed fall into the dark space of the bucket.

24 September
My friend Abdul tells me of an incredible large-scale rewilding project taking place in north-eastern Siberia. It is known as Pleistocene Park – born of the brilliant mind of Russian scientist Sergey Zimov, who is looking to rewild enormous areas of what were once mammoth territory; steppe ecosystems lost since the last Ice Age as humans migrated into the Siberian Arctic and killed off the large herbivores. Zimov and his son Nikita aim to repopulate the region by bringing in bison, reindeer, wild horses, wolves and tigers.

When there were grazing animals, they would clear the snow to reach the grasses below. This in turn meant that when colder weathers came in winter, the land froze hard again, maintaining the permafrost. Without the herds of herbivores, snow lies undisturbed, insulating the land beneath, which means these wild icelands of the north are beginning to thaw. As they do, vast quantities of methane and carbon dioxide are released into the atmosphere.

The sheer size of the area is hard to imagine. It is a great, vital venture to halt the melting of millions of acres of permafrost.

We need people like Sergey Zimov. We need minds that can see through global climate change and radically shift ways of being. They give us all hope. Zimov is not acting to make money through rewilding, to become rich. He is acting to save the world.

27 September
An autumn sowing.

I have scythed a further section of meadow and scarified the ground. Now, I scatter a selection of seeds I have collected in little brown bags.

> Figwort
> Devil's-bit scabious
> Cornflower
> Ox-eye daisies
> Corncockle
> Dropwort
> Yellow rattle

I wish them luck as I step gently on the seeds, pressing them into the earth.

28 September
The conversation with Hugh Somerleyton has really stuck with me. That notion of democratising nature recovery rings around my thoughts. It makes such sense.

I now know that the best way forward is the establishment of government-funded, large-scale rewilding projects overseen by ecological experts like Steve Carver, where we might even witness exciting reintroductions of apex species like wolves and lynx into the landscape. But on other, far smaller levels, we all have to be engaged in renaturing. We must all be part of the movement.

Most of us humans live in cities, so it is in urban spaces, too, where there needs to be an urgent wave of nature recovery that people can see and be a part of.

'We're about greening up more urban space,' Errol Fernandes states. 'All the gardens in London, for example, would take up a huge amount of hectares.'

On the other side of the screen, I'm nodding.

That is exactly what democratising nature recovery means – seeing the collective environmental impact of thousands upon thousands of individual efforts.

'Perhaps we need to see *ourselves* as keystone species,' he says. 'We are agents for disturbance in urban environments. If we can bend our thinking a little more and see that our presence in a place can also be beneficial for wildlife, making space for nature – biodiverse planting, etc. – then we start to engage people more rather than this sense of climate guilt we all seem to be burdened with, the gloom and doom and destruction.'

Errol is sat in his office, taking a moment out from overseeing the sixteen acres of gardens he's in charge of at the Horniman Museum in Forest Hill, South-East London. I first heard Errol speak only a few weeks before at the 'Rewilding the Mind' conference in a discussion of rewilding and wellbeing. Now I have the chance to chat with him. I stepped out of the field only a moment ago – changing my mud-worn field outfit for my work-from-home office wear.

'I've struggled to watch the new David Attenborough series,' he confesses. 'I want to. My kids want to. But . . .'

'Absolutely,' I agree.

'The trouble is that though the series is amazing viewing, the emotional journey is so full of eco-anxiety, eco-doom. There seems to be such a sense of individual helplessness.'

He speaks so well. I already feel a connection to Errol. We've realised we both grew up in the suburban wastelands of West London, sharing a love of digging up bottles on Victorian dumps as children and running wild on brownfield sites. He's very easy to talk to.

'In Britain, it's hard to do rewilding properly because we don't have the space. Much of the land we do have is owned by the Crown Estate or a minute number of extremely wealthy landowners.'

Errol nods back. 'When we talk about wilding or rewilding, it really does feel like we're talking about the great and the good landowners, like we're articulating something that is the preserve of the privileged.'

'Exactly,' I agree.

'That really bothers me. Aligning with that doesn't make any sense,' Errol states. 'That is not what I am doing here. We sit within a community, within the borough of Lewisham. We're striving to open our doors to a cross-section of our local community. For us, it's about being good stewards and custodians. It's about the people who come here and why they come – educating and trying to influence change within society for the better.'

The Horniman Museum is certainly doing that. They were recently named Museum of the Year. Close to a million visitors have gone through their doors and gardens in the last year. Errol tells me of a previous role working at the Kenwood Estate, on the other side of the city, at Hampstead Heath, North London. The gardens there had been forged into their present form by Humphry Repton in the late eighteenth century – one of those key names in landscape design who oversaw large-scale renovation on many of England's country estates. By the time Errol got to Kenwood, the area was rapidly reverting back to a wilder state.

'The sheep that had been keeping down the bramble, preventing the whole place returning to forest, had long gone. I replaced them with sixty volunteers, who each week would come religiously with mattocks to dig up the bramble.'

'Which is a tough job,' I say, remembering my own efforts.

Errol smiles. 'That they love dearly.'

I laugh.

'They were amazing!' he continues. 'We would clear six tons of bramble a week.'

I think of my ongoing muddy battles with the wall of bramble in the field.

What I really want to ask Errol about is one particular project that he is engaged with – the construction of what is known as a micro-forest. He explains how when he moved to the Horniman Museum he was keen to develop a corridor of land neighbouring the South Circular Road as just that – a section of micro-forest centred around the pioneering work of Japanese botanist Akira Miyawaki.

The Miyawaki Method, as it is known, involves planting native species of trees tightly together after ensuring they have loose, rejuvenated soils full of organic matter to grow in. Selecting which trees to grow as a micro-forest is made after careful investigation of what Miyawaki called the potential natural vegetation, or PNV, of any piece of land based on a variety of factors including topography, climate and soil conditions.

'Obviously, planting natives is important because they've evolved symbiotic relationships with many of the species around,' says Errol, explaining his thinking in constructing the micro-forest at the Horniman Museum.

'However, we also had the opportunity to experiment and explore and learn about planting some trees that we could hypothesise to be able to cope better with more sporadic rainfall and increased temperatures.'

Essentially, he was looking to future-proof the project against climate change.

'We planted a backbone of British natives, and feathered in some more unusual species from further afield.'

But the key factor for Errol was to plant and then be 'hands-off'. He wasn't seeking to design and create a stylised human-forged forest. Instead, it was more about seeing how the sapling trees evolved.

'Of course, we're not completely hands-off. We've been irrigating. We're not hands-off, because we planted the trees! We're kick-starting something and then we want to watch and wait and see.'

That learning aspect to the process is central to the project, fitting to the educational framing of the museum ethos. Yet there are some things that *are* known about micro-forests.

'We know that micro-forest planting sequesters much more carbon than more traditional planting because of the way the trees grow so vigorously,' Errol states.

'We know that we will get a closed canopy in fifteen to twenty years that would otherwise take a hundred and fifty to two hundred years.'

The micro-forest is being grown right beside one of London's busiest roads.

'We know that many of the trees we planted will capture particulate pollution in their leaves and that a depth of ten metres or more of forest will act as a sound-buffer to the traffic noise.'

The 30 different species and 900 trees of the micro-forest will also help to drive up biodiversity on the museum site.

'The gardens are an oasis within South London, right beside a busy road. We're working to create a place that connects people with nature. That's probably the most important job we can do.'

Errol explains how his previous profession was as a psychotherapist.

'Our connection with nature seems broken and disparate. If we are connected with nature, we are healthier – we've got the stats on it – in relation to wellbeing, to exercise . . .'

RENATURING

We've already joked about the similarity of our backgrounds in suburban West London.

'We didn't have Netflix, we didn't have iPhones, we didn't have the internet,' I say. 'For me, the idea of being inside was boring, so I was always outside.'

The world is very different now. Connecting to nature isn't always that easy. Places such as the Horniman Museum can help.

'We're providing a wilderness space that *feels* like a wild place,' says Errol.

His passion for his work, for allowing this vital natural connection for the children of South London is so obvious. He's also a busy man. Our time is up. We say our goodbyes, and head our own ways.

I change back into my muddy field gear and head out.

On 'Rewilding' a Window Box

That question of whether you can 'rewild' a window box is in essence a question of definitions, about how you see the notion of rewilding.

True rewilding projects need space – thousands of acres of space. They need big scale so as to allow the presence of top predators that will bring top-down trophic control over the region. Think wolves and lynx roaming a vast wilderness. Obviously, window-box rewilding can't offer that.

Yet there is a problem in a country like Britain, where so much of the land is owned by a wealthy elite. How can the vast majority of the people be an active part of the nature restoration of the country? What about those people who live in urban areas? What of those without a garden, without any outdoor space but a windowsill? That's where the idea of micro-interventions towards nature recovery springs to life.

Take an iconic tower block like Trellick Tower in Westbourne Park, just north of Notting Hill in West London. There are 217 flats. If each flat occupant had a window box six inches by eighteen inches, and each window box were planted with pollinator-rich wildflowers, then collectively they would form eighteen square yards of wildflower meadow. In fact, each flat at Trellick Tower has a south-facing balcony, so you could ask each occupant to plant five or ten times that window-box space. Suddenly you're looking at over four acres of wildflower meadows spreading across a tower block of balconies. Then perhaps get someone to put a beehive or two on the roof? An active ecosystem is starting to be formed.

There are obviously issues with seeing this as rewilding because it really *isn't* rewilding. But such active engagement in renaturing a local landscape – even if that means window boxes – allows urban populations to be directly involved, to be engaged in nature restoration. That's the key. That's why getting people to plant wildflowers in window boxes is a great idea, whatever you chose to call it.

ENDINGS

'The world is mud-luscious and puddle-wonderful.'
                              e e cummings

It feels as though I have been on a long, winding voyage to far-distant lands. Yet I am here still – stood in the field where once I sat on a machine and mowed some rough form of cricket pitch to play upon with my friends; where now I mow patches of wildflower meadow with a scythe.

Summer has become autumn, then winter. Still I am here, working away in the field.

Now, the bramble has been cut back and the wildflower meadows have been sown. There is still the leggy, tired blackthorn hedge beside the green lane that needs to be cut down and laid out as nesting sites for songbirds in the spring. There will continue to be bramble to cut back and in time there will be meadows to mow. There will always be a need for intervention in an ecological restoration project of this size.

Renaturing the field will go on and on.

One saga has gone on for long enough. From the very first days of renaturing the field, a pond was always a key part of that vision. But the pond is yet to be dug. I have failed to dig it with my own spade, so I will get in a digger and imagine the churned ground of the caterpillar tracks has been due to a horde of aurochs that have passed this way. Over the last few months, I've worked away in snow, rain and sun to make the field a better place for nature. I've even saved an apple tree from girdling. But there is still no pond. It is the last piece of the project.

A year or so ago, I spoke to my friend Mike Penfold about who was the best person to do the digging. He'd grown up a couple of villages over. Mad Mark or Digger Dave were the two candidates he recommended.

And another year had somehow gone by in the usual flurry.

Now, I will seek the moment.

I text Mike:

> Could you let me have Digger Dave's number? Reckon he's the man for making a pond, eh?

His response pings back.

> He is the leading pond creator in the whole of Essex. Some say he can hold his breath underwater for two hours. Will text it over.

I laugh, thank Mike, and like magic Dave Cowling's contact details appear. I reply, 'Half man, half frog', and give Dave a call. It has been a number of years since I last spoke to him. A deep voice answers.

'Dave?'

'Yes?'

'Oh, Mike Penfold gave me your number . . .'

'Oh, yeah.'

'Actually, I used to teach your daughters,' I begin.

'Ah . . .' says the voice breaking to a lighter tone, 'Mr Canton.'

I laugh.

We catch up on how life has been treating him and his daughters Tiffany and Bryony before turning to the matter in hand – the digging of a pond.

'Do you know what the ground's like?' Dave asks.

'There's solid clay from about six inches down.'

'Right,' says Dave.

'I was wondering about puddling the pond,' I venture.

Dave explains. 'Puddling is if you need to bring in clay to line the pond. Sounds like you'll be alright for clay.'

'Reckon so,' I say, remembering what Ashley Cooper had told me back in June.

Dave agrees to head over in a couple of days to take a look.

'I have been to yours once before but it was a while back,' he says.

My mind is a blank. Had he?

'Down a green lane, isn't it?' Dave asks.

'Yeah, that's it.'

I give Dave more detailed directions for where I live and we say our goodbyes for the time being. But I'm left somewhat bemused.

When had he been over before?

It wasn't until the next day that I remembered – Dave had actually been the first person to cut the field down after I'd moved into the cottage. It had been over twenty years ago. There was a delightfully pleasing symmetry, it now struck me, in the way that it would also be Dave who would carry out the last act of serious intervention on the field. After the construction of the pond, there would be no more tractors or diggers – it would simply be a matter of managing the place: scything, bramble-cutting, occasional seed-collecting and scattering.

When Dave arrives a couple of days later, we unpick the moment of his last visit.

'I came and did some topping,' he explains.

'That's when I first bought the field,' I say.

'It was ten, fifteen years ago,' adds Dave. 'Could be more than that . . .'

'I think it was more,' I say.

For a second or two, we both think back to that time.

'I bought it in May 2000 and I think you came over that June or something,' I say slowly. 'The grass was . . .'

I stick my arm out at waist height.

'It was,' agrees Dave. 'It was tall.'

'I'd just moved up from London. I didn't have a clue how to cut it down,' I confess.

We both laugh. It was a lifetime ago. Then it had been all about trying to mow the meadow grasses down as much as possible. The plan had been to turn an abandoned patch of ground into a cricket pitch for the match promised to my London friends for later that summer. I had gone out and bought an aged sit-on mower, but it had proved useless in the three-foot-high pasture that the field had turned into. Fortunately, one sunny day Dave had arrived on a tractor and soon had topped the wilderness down to something vaguely possible to play cricket on.

The field had been on quite an adventure since that first cut by Dave twenty years back. Now, we stroll over towards the site where the pond is to go and arrive at the small section of land I have dug.

'As you can see, I made a start,' I explain. 'I wanted to dig it by hand.'

I smile.

'Then I realised that wasn't going to happen. So it's just sat like this . . .'

'Mike mentioned it to me one night in the King's Head in Pebmarsh,' recalled Dave. 'I should think it was a year ago.'

We stand for a few seconds gazing at the ground around us.

'So what sort of size do you want?' he asks.

It is a good question. I look back down at the damp patch of earth before us.

'From the edge of the dug patch to the brambles, or so . . .' I say, pointing to our left. 'As deep as possible.'

They are hardly precise directions.

'OK, sure. You plot the shape that you want,' suggests Dave. 'I'll dig the area, take the top soil off and dig the sub-soil to the depth you want. But what do you want to do with the stuff that comes out? There will be quite a lot.'

I laugh.

'Yeah, Selfie said there would be.'

'It bulks up by about a third,' adds Dave.

We agree that he will stick it on top of the mass of bramble to the west end of the pond, to form some kind of bank.

'That will save me a job,' I say, thinking of the hassle of cutting back that bramble and mattocking out the roots. 'The only thing I'm concerned about is that the hole holds water,' I add.

'That *is* the thing,' agrees Dave with a smile. 'The land around here is very variable. You can be digging in clay one minute and the next you're in a seam of ballast. You won't know until you've dug it. Depending on the clay that comes out of it, if there are ballasty bits you can puddle that in.'

There's that word again. To puddle a pond.

It will be so much better to have a natural pond, lined with clay rather than with a large and expensive plastic liner.

'Anyhow, if you can roughly plot out the size and shape,' he says, 'I can start next week.'

Dave will return at the weekend with his digger on the lorry and leave it here. Then he will begin digging next Tuesday. All is set.

A few days later, over the weekend, a three-ton orange digger appears in the far end of the field. With sticks, I mark out a rough shape for the pond some thirty feet long and twenty feet across at the widest point. On the Monday, rain – which has been absent for much of the month – begins to fall. By Tuesday, when Dave arrives, a steady drizzle has settled in. It isn't good digging weather. We agree to try again the next day.

On the Wednesday, I return from dropping Eva at the school bus stop to the steady soft hum of an engine coming from the field. The rain of the last few days has stopped. Sunshine is peering through grey clouds. I head over and see Dave in the digger with his radio on. A large hole has appeared in the field. He shuts down the engine as I arrive.

'Alright, Dave?' I say.

'How are ya?'

'Good,' I say. 'Is it OK?'

'It's a bit sticky, but it's doable.'

'Looks amazing,' I add. 'I'm so glad you're doing this!'

'Rather than you?'

'Yeah. Took me about a month to just do that bit.' I point to the foot-deep patch beside us. 'Then my knees went.'

Dave laughs.

I leave him to it. He is in his element out there, settled in the cab of his digger with Radio Caroline to keep him company. Halfway back across the field, I glance around to see the bucket plunging into the ground. I can just make out the sound of music beneath the steady rumble of the engine.

By the time I return to the field a couple of hours later, there is a deep scar thirty feet long.

'Wow,' I murmur under the growl of the digger.

Dave sees me and shuts the engine down.

'Certainly getting there,' I say.

'So what you've got to decide is how deep you want it?'

'Sure,' I say. I step carefully into the hole. 'Is it clayie enough, do you think?'

Dave hesitates.

'It's not the sort of clay you'd build a lake out of. It's a bit loamy.'

The subtle variance in the subsoil is easy to make out from within the hole. We both begin to handle the clay. The paler, loamier version doesn't compact up as well as the rather darker form.

'So how do you go about puddling it?' I ask, squeezing a cold clod of clay in my hand.

'You track it in,' explains Dave.

Using the digger, he will bring in bucketfuls of the good clay and then drive up and down over it to form a base layer. With luck, that will be watertight. Only it isn't just luck. I know that. It is also skill and experience – a deep knowledge of working with clay, and with diggers, years of forging farm ponds for miles around.

'I'll go down deeper and you decide if you want the pond extending out further,' he says. 'Of course, the deeper you go, the steeper the sides.'

Having a deep section is crucial for some of the creatures who will hopefully arrive in due course and make this their home. Yet the pond also needs to have easy access.

'Here you have your shallows,' points out Dave.

There is a clean, smooth gradient.

'That looks lovely,' I say. 'So the frogs and whatever can get in and out.'

We stand a few seconds.

'It's a decent size,' I add.

'It is,' agrees Dave.

I leave him to get on.

It is after lunch when I return. It is obvious that Dave has been busy. The hole has been smoothed off and now is over five foot down at its deepest. It all looks very pond-like.

'You're getting on,' I joke.

He laughs.

'Looks like better clay as well,' I say, pointing to the middle.

'It is,' agrees Dave. 'Because under that it got very hoggardy. So I dug it out and put better stuff back in there.'

'Ah. Puddling!'

'Yeah, sort of,' he says. He laughs again. 'I've blended the clay in, but really, you should use compaction equipment, rollers and the like to knit it all together.'

It is good to get the final word on puddling – from a proper pond puddler.

'So what do you do if you're digging a pond and you hit a bit that's got no clay at all?'

'You have to get clay and bring it in.'

Kind of obvious, I guess.

'Bet that's a right faff,' I say.

'It is,' agrees Dave. 'It's an expensive faff, too.'

He smiles.

'Ideally, what you want is London blue clay,' he explains. 'It's got no stone, nothing in it at all. It's a beautiful blue. That's what they used to make bricks out of. That is the best thing, but there's none around this area. I've been on jobs when people have brought clay in and it hasn't actually worked. The clay they've brought in is too loamy and the water soaks through it.'

I look at the big hole before us.

'You'll just have to see what happens here,' he says.

It's not exactly a cast-iron guarantee.

'Sure,' I say. 'And if it doesn't work . . .?'

There's a slight second of hesitation. Dave looks at me straight-faced.

ENDINGS

'I'll give you a bell,' I say, and we both laugh out loud.

In truth, I have faith.

'It should be OK,' says Dave after a moment.

I leave him to his lunch.

When I return later in the day, he has smoothed off the edges of the pond and formed a raised terrace to the west some three feet high. It's been made from the clay not used to puddle the pond. Dave sits in the cab, the worn material of the seat evidence of his time at the wheel. The digger arm swings around, the bucket brushing against the ground like a giant's hand gently smoothing the clay. In less than a day, he has dug out a tear-shaped hole some thirty feet long by twenty feet across and over five feet deep in the middle. But what seems more remarkable is how this great gouge in the field looks. It is so tidy, so neat.

'That is so impressive,' I say, my voice not audible above the engine noise.

Dave shuts the digger down.

'Brilliant,' I say.

He steps from the cab.

'You happy with that?'

'Yeah, I'm happy.'

'Good,' he says. 'Then I'm happy.'

The pond is dug. Or rather, a large hole is dug.

So what do you do when you've suddenly got a large muddy hole in a field? You pray for rain. And you pray that the water will stay and become a puddle and that gradually the hole will become something resembling a pond.

The next morning, I step out to the field and walk eagerly over along the winding deer path through the meadows. It has rained in the night – not bucketfuls, but enough. This is the test. Dave's orange digger still sits silently at one end of the hole. I weave through the young oaks and come to the edge of the pond. The sides look even smoother than yesterday, the clay a pale sandy colour. There, in the bottom, lies a shallow pool of rainwater.

It is a good sign. And over the next couple of weeks, the rain continues to fall and that muddy puddle grows and grows. A few oak leaves gather on the surface. Perhaps inevitably, I seek out ways to help the rising waters. I hit upon the idea of wheelbarrowing in water from the water butts around the cottage, which are overflowing with lovely rainwater. The path through the field gets muddier and more waterlogged. The barrow slides and slips as I go, spilling its cargo onto already sodden ground. I watch as the level edges up the shallow slopes Dave has sculpted with the digger bucket.

In a moment of clarity, standing by the hole in the dusk of an early evening, empty wheelbarrow in hand, wet from still-falling rain, I think how my wanting to step in, to help, to manage the process of the hole turning to a pond, is really so much a lesson in renaturing. Intervene in the landscape and then step back. Be conscious of when to initiate change in domesticated lands and then let natural processes take over to make them wild again.

'Just leave it,' agrees my old friend Mark the next night, sat on the floor of his cottage in Kersey. 'It'll fill in good time.'

He is right. Of course, he is. I can even hear the rain falling outside as he speaks.

Yet it is intriguing to feel that urge to be involved, to know that urgent desire to oversee the regeneration of the field rather than merely letting natural processes do their work. For the last few thousands of years, we have evolved into beings that want to manage, to domesticate the world around us. But for far longer before then, we knew when to act effectively, when to step in and scatter some seed, when to clear land. On small patches of earth like the field, it can be good to step in and help ensure the best environment exists for all the living beings in that space. Yet we also need to remember when to relinquish control, when to halt that desire to do, to act, and merely let nature be.

As I'd stood by the muddy hole in the rain, something else had been becoming clear, seeping into my conscious thoughts. The creation of the pond was the culmination of the plan that had begun more than a decade ago when I had stopped cutting the outer edges of the field and the first oak saplings had begun to rise from jay-sown acorns hidden far beneath the meadow grasses.

Except, of course, there would be no actual end point – the evolution, the natural regeneration of this patch of earth would continue for years. There would still be the need to scythe the meadows each autumn and to scarify the ground, to act as though a herd of bison or aurochs had stumbled through this patch of earth on their annual migration, to imitate the presence of those creatures lost to this land.

Rain would fall.

The muddy hole would gradually turn into a pond.

The sun would shine.

Wildflowers and meadows would rise from the earth once more.

Birds would sing.

The young oaks would grow ever closer to the sky.

All would be well.

ACKNOWLEDGEMENTS

To the Memory of Ray Davis, Neil Guest and Jonathan Jukes

Renaturing emerged from a desire to see the field behind my cottage grow wilder, and evolved as I learnt more of the true nature of rewilding and the means by which we humans can all be a compassionate, caring element in the process of restoring the natural world. Along the way, I sought the expertise and voices of many to help tell this tale. The following good people were especially generous: Anna Beames, Noah Bennett, Stephen Briggs, Richard Brown, Julia Boulton, Glenn Buckingham, Steve Carver, Ashley Cooper, Dave Cowling, Ray Davis, Hazel Draper, Jack Durant, Errol Fernandes, Chris and Jude Gibson, Paul Gwynne, Polly Lavender, Richard Mabey, Mark Mansfield, Frances Mount, Annie Randall, Ben and Louise Rees, Mark Self, Guy Shrubsole, Natalie Singleton, Hugh Somerleyton and Darren Tansley. My thanks to them all for the time, knowledge and kindness they offered.

    A small but beautiful collection of individuals has been absolutely key to my scribblings and thoughts actually becoming this book: my sister Helen Canton, my partner Madeleine Last, my tutor Peter Hulme, my editors Simon Thorogood and Claire Reiderman, and my agent Jessica Woollard. I have been so grateful for their guidance and love as the work has come together from the earliest stages to the very last proofs.

Many others offered wise words and thoughts or supported me in various ways to find the time and space needed to write *Renaturing*. They, too, have all played their parts and I thank them: Abdulkareem Atteh, Helen Bleck, Ronald Blythe, Jane and Bruce Boulden, my children Eva, Molly and Joe Canton, my mother Margaret Canton, Ben Castell, David and Mandy Charleston, Yalda Davis, Katie Dawson, Matthew De Abaitua, Hannah Elliman, Neil Gant, Ros Green, Betty and Colin Greenaway, Neil and Lynda Guest, Tony and Cindy Hilling, Liz Kuti, Juliet Lockhart, Ginny and Matt Mackman, Sara Maitland, Lucy Manthorpe, Adrian May, Chris McCully, Ellie Mead, Andy Papps, Mike Penfold, Craig Perry, Holly Pester, Jordan Savage, Diane Smith, the staff at the British Library and the Albert Sloman Library at the University of Essex, Jack Stimpson and Phil Terry.

To all those at Canongate who have worked so brilliantly on the book I also offer my gratitude, especially Leila Cruickshank, Jenny Fry, Jamie Norman and Lucy Zhou.

There are also those whose names I cannot recall or never knew that I have met along the journey who have helped in their own ways towards making *Renaturing* the book it is – from the fine folk chatted to at talks and festivals, to those passed in lanes and pathways who have exchanged a few words on a country walk. To them, too, I send my thanks.

ENDNOTES

All online material last accessed 18 April 2024.

1 Leo Tolstoy, *Anna Karenin* [1878], translated by Constance Garnett (1901), Part 3, chapters 4 and 5, available at: gutenberg.org/cache/epub/1399/pg1399-images.html#chap03.
2 Andrew C. Scott, *Burning Planet: The Story of Fire through Time* (Oxford: Oxford University Press, 2018), p. 150.
3 Further reading:
   'Invasive Species Week: What Can I Do?', GB Non-native species secretariat website, available at: nonnativespecies.org/what-can-i-do/invasive-species-week/.
   'Species reintroduction and conservation', RSPB website, available at: rspb.org.uk/helping-nature/what-we-do/influence-government-and-business/species-reintroduction-and-conservation.
   'Tumbleweeds: the fastest plant invasion in the USA's history', Natural History Museum website, available at: nhm.ac.uk/discover/tumbleweeds-fastest-plant-invasion-in-usa-history.html.
   'Invasive Non-Native Species', U. S. Environmental Protection Agency website, available at: epa.gov/watershedacademy/invasive-non-native-species.
   'Invasive non-native species (UK) – Grey squirrel', Inside Ecology website, available at: insideecology.com/2017/10/12/invasive-non-native-species-uk-grey-squirrel/.
4 William Cobbett, *The Woodlands* (1825), quoted in Richard Mabey, *Flora Britannica* (London: Sinclair-Stevenson, 1996), pp. 197–8.
5 Alexis Nikole Nelson, @blackforager, Instagram post, 23 May 2022.
6 Jack Lindsay, *The Discovery of Britain* (London: The Merlin Press, 1958), p. 25.

ENDNOTES

7   Further reading:
    Big or small, ponds for all, Wild About Gardens booklet, accessible from www.wildaboutgardens.org.uk/.
    'Wildlife ponds', at www.rhs.org.uk/ponds/wildlife-ponds.
    Sylvia Myers and Lisa Hendry, 'Pond life: facts about pond habitats, plants and animals', Natural History Museum website, available at nhm.ac.uk/discover/pond-life-facts-about-habitats-plants-animals.html.
8   Ronald Blythe, *Outsiders* (2008), quoted in Leif Bersweden, *Where the Wild Flowers Grow* (London: Hodder & Stoughton, 2022), p. 10. Copyright © Ronald Blythe, 2008. Used by kind permission of the Ronald Blythe Estate, courtesy of Ian Collins.
9   Isabella Tree, *Wilding* (London: Picador, 2018), p. 67.
10 David T. Schwartz, 'European Experiments in Rewilding: Oostvaardersplassen', on Rewilding.org website, available at: rewilding.org/european-experiments-in-rewilding-oostvaardersplassen/;
    Patrick Barkham, 'Dutch rewilding experiment sparks backlash as thousands of animals starve', on *Guardian* website, available at: theguardian.com/environment/2018/apr/27/dutch-rewilding-experiment-backfires-as-thousands-of-animals-starve.
11 'Defining Rewilding', on Rewilding Britain website, available at: rewildingbritain.org.uk/why-rewild/what-is-rewilding/an-introduction-to-rewilding/defining-rewilding.
12 Further reading:
    Dolly Jørgensen, 'Rethinking rewilding', *Geoforum* 65 (2015), pp. 482–8.
    'An introduction to rewilding', *Oxford Advanced Learner's Dictionary*, available at: learningenglishwithoxford.com/2021/05/28/introduction-rewilding-oxford-dictionary/.
    'Defining Rewilding', Rewilding Britain, available at: rewildingbritain.org.uk/explore-rewilding/what-is-rewilding/defining-rewilding.
13 See Mark Avery's invaluable work on Walshaw Moor in his Wuthering Moors archive: markavery.info/archived-blog-posts/walshaw-moor-wuthering-moors-archive/.
14 Isabella Tree and Charlie Burrell, *The Book of Wilding: A Practical Guide to Rewilding Big and Small* (London: Bloomsbury, 2023), pp. 4 and 7.
15 See: Katie Pavid, 'What is biodiversity and why does its loss matter?', Natural History Museum website, available at: nhm.ac.uk/discover/what-is-biodiversity.html; 'Biodiversity – our strongest natural defense

against climate change', UN Climate Change website, available at: un.org/en/climatechange/science/climate-issues/biodiversity.
16 Dave Goulson, *The Garden Jungle: or Gardening to Save the Planet* (London: Vintage, 2019), p. 226.
17 Chris Packham, Patrick Barkham and Rob Macfarlane, 'A People's Manifesto for Wildlife' (2018), available at: chrispackham.co.uk/wp-content/uploads/Peoples-Manifesto-Download.pdf.
18 Geoffrey Grigson, *The English Year: From Diaries and Letters* (Oxford: Oxford University Press, 1967), p. 97.
19 Cain Blythe and Paul Jepson, *Rewilding: The Radical New Science of Ecological Recovery* (London: Icon Books, 2020), p. 7.
20 See Dennis M. Hansen, 'Non-native megaherbivores: the case for novel function to manage plant invasions on islands', *AoB PLANTS* 7 (18 July 2015), pp. 1–11, available at: doi.org/10.1093/aobpla/plv085
21 Community of Arran Seabed Trust (COAST) to give them their proper name.
22 '"Pretty damn cool": Ellie Goulding on rewilding as a cure for our planet – and our mental health', Wild World: A Rewilding Special, *The Guardian*, 25 June 2022, available at: theguardian.com/environment/2022/jun/23/ellie-goulding-on-rewilding-as-a-cure-for-our-planet-and-our-mental-health-aoe.
23 'Finding Wild Food with @Blackforager', *Extra Spicy* podcast, 22 February 2021, available at: podcasts.apple.com/gb/podcast/finding-wild-food-with-blackforager/id1517318479?i=1000510137418.
24 Brooke Mitchell-Norman, 'How Do We Rewild the World?', *Rewildology* podcast 69, 31 March 2022, available at: rewildology.com/2022/03/31/how-do-we-rewild-the-world-episode-69.
25 See also y2y.net for information on the Yellowstone to Yukon Conservation Initiative.
26 'Are ESG credentials compatible with landscape-scale habitat restoration schemes?', webinar on Savills UK website. Available at: savills.co.uk/landing-pages/are-esg-credentials-compatible-with-landscape-scale-habitat-restoration-schemes-.aspx.